INFRARED METHODS FOR GASEOUS MEASUREMENTS

OPTICAL ENGINEERING

Series Editor

Brian J. Thompson
William F. May Professor of Engineering
and Provost, University of Rochester
Rochester, New York

Laser Engineering Editor: **Peter K. Cheo**
United Technologies Research Center
Hartford, Connecticut

1. Electron and Ion Microscopy and Microanalysis: Principles and Applications, *by Lawrence E. Murr*
2. Acousto-Optic Signal Processing: Theory and Implementation, *edited by Norman J. Berg and John N. Lee*
3. Electro-Optic and Acousto-Optic Scanning and Deflection, *by Milton Gottlieb, Clive L. M. Ireland, and John Martin Ley*
4. Single-Mode Fiber Optics: Principles and Applications, *by Luc B. Jeunhomme*
5. Pulse Code Formats for Fiber Optical Data Communication: Basic Principles and Applications, *by David J. Morris*
6. Optical Materials: An Introduction to Selection and Application, *by Solomon Musikant*
7. Infrared Methods for Gaseous Measurements: Theory and Practice, *edited by Joda Wormhoudt*

Other Volumes in Preparation

INFRARED METHODS FOR GASEOUS MEASUREMENTS

THEORY AND PRACTICE

Edited by

Joda Wormhoudt
Aerodyne Research, Inc.
The Research Center at Manning Park
Billerica, Massachusetts

MARCEL DEKKER, INC. New York and Basel

Library of Congress Cataloging in Publication Data

Main entry under title:

Infrared methods for gaseous measurements.

(Optical engineering ; v. 7)
1. Flue gases--Measurement--Remote sensing.
2. Infrared radiation. 3. Infrared spectroscopy.
4. Molecular spectroscopy. I. Wormhoudt, Joda,
[date]. II. Series.
TD885.I54 1985 628.5'32'0287 85-7013
ISBN 0-8247-7429-9

MARCEL DEKKER, INC.
270 Madison Avenue, New York, New York 10016

Current printing (last digit):
10 9 8 7 6 5 4 3 2 1

PRINTED IN THE UNITED STATES OF AMERICA

About the Series

Optical science, engineering, and technology have grown rapidly in the last decade so that today optical engineering has emerged as an important discipline in its own right. This series is devoted to discussing topics in optical engineering at a level that will be useful to those working in the field or attempting to design systems that are based on optical techniques or that have significant optical subsystems. The philosophy is not to provide detailed monographs on narrow subject areas but to deal with the material at a level that makes it immediately useful to the practicing scientist and engineer. These are not research monographs, although we expect that workers in optical research will find them extremely valuable.

Volumes in this series cover those topics that have been a part of the rapid expansion of optical engineering. The developments that have led to this expansion include the laser and its many commercial and industrial applications, the new optical materials, gradient index optics, electro- and acousto-optics, fiber optics and communications, optical computing and pattern recognition, optical data reading, recording and storage, biomedical instrumentation, industrial robotics, integrated optics, infrared and ultraviolet systems, etc. Since the optical industry is currently one of the major growth industries this list will surely become even more extensive.

Brian J. Thompson
University of Rochester
Rochester, New York

Preface

It is not too much to ask that a preface describe the contents of the book it precedes and indicate the audience to which it is addressed. In the case of this book, it *was* too much to ask of a title—the few words on the cover require a few more by way of explanation. Measurement techniques using infrared radiation, in both theory and practice, are indeed the subject here. Beyond that, the first additional comment that should be made is that the applications typically envisioned here are to industrial flames and stack exhaust plumes. The adjective "industrial" is in part only a shorthand way of distinguishing these applications from other areas in which infrared technology has been important, such as in the analytical laboratory, in the study of the upper atmosphere, and in the remote sensing of properties of the earth from space.

In fact, most of the material in this book was developed through the efforts of various government agencies, with the military contributing its support along a broad front from the most basic science to a wide variety of specialized systems, and other agencies such as NASA and the Environmental Protection Agency pursuing their own charters. Nevertheless, the work they began should now find increasingly practical applications to industrial problems, including the determination of temperatures and chemical species concentrations in flames, such as burnoff flares and furnace flames, and in exhaust gases and stack plumes, and the measurement of trace gas concentrations and other physical properties both in the immediate vicinity of industrial sources and far away from them. It is a major thrust of this book to aid in effecting a technology transfer to new applications, especially those in private industry.

A second qualification is that the infrared measurement techniques discussed here are in a sense "passive." Most of the book deals with a truly passive technique – detection of emission from hot gases. Where the more informative emission–absorption method is used or, in observing the properties of the ambient atmosphere, where absorption rather than emission measurements are appropriate, a (continuum) source of infrared radiation is required. These techniques cannot in fact be construed to be passive except in comparison to the most active of infrared devices, the laser, about whose applications a great many books have already been written.

A third point to be made here is that a main focus of the book is on high resolution spectra, in which individual molecular vibrational–rotational lines can be seen. In some ways, this makes them easier to use—higher peak intensities make it easier to detect emission lines, characteristic patterns of lines make it easier to measure concentrations when there are interfering bands, and a high resolution spectrum typically offers several ways to obtain a temperature measurement. On the other hand, to accurately extract any of this wealth of information requires expertise in a wide range of disciplines, including fluid dynamics, molecular physics, radiation transport, and often chemical kinetics in the gas source, atmospheric transmission, then optics and electronics in the detection system. This convergence of a number of areas of knowledge makes this a challenging and exciting field of investigation. With all of the techniques discussed here depending on a knowledge of the line structure of gas molecule absorption and emission, the term "high resolution" could well have been another candidate for inclusion in the title. However, the actual emission measurements may instead be done in selected filter bandpasses, while the nondispersive absorption methods discussed here rely on correlation with a reference cell of the gas to be identified, rather than on the resolving power of a spectrometer or interferometer. Nevertheless, the spectral nature of infrared radiative properties of gases is the theoretical basis for this book. Its purpose is to introduce that theory, as well as some practical aspects, to a wider audience.

As to the audience, this volume is especially directed to people who want to use high resolution infrared spectral features to carry out automatic, remote monitoring and control of hot gas systems like flames and stack exhaust gases, and of cool gas systems like the atmosphere. Fourier transform and nondispersive infrared equipment and the incorporation of microcomputers into equipment have all made great progress in recent years, and a number of people have begun to think about their applications. However, connection between the observed radiation and the desired quantities such as temperature and species concentration involves a knowledge of basic physics and chemistry as well as experience in the practical details of infrared technology. We can use that knowledge to calculate system performance. This volume includes qualitative and quantitative introductions to the physics involved in using remote observations to collect data, as well as examples of these data. The individual chapters provide reviews of the work in their specific areas, supply tutorial or educational information, and present exemplary data and calculations. Despite varied subject matter and approaches, they share the objective of communicating to instrument developers and potential industrial users the basic physical principles underlying infrared measurement techniques, thereby allowing them to judge their usefulness and range of application.

Chapter 1, by John Conant, Bill Herget and me, compares burnoff flare and stack plume emission spectra taken by Herget with parametric calculations performed with equally high spectral resolution. Similarities to the observed data were obtained in what was only a simple calculational exercise, and there is no doubt in our minds that further comparison and analysis can result in very realistic model predictions, which can form the basis for converting observation into information.

This first chapter poses several problems, each deserving at least a chapter of its own, and some do receive more in-depth treatment in this book. For one, the contribution from Bill Prengle's group, in Chapter 2, gives some details on the fluid mechanics considerations that must underlie any first-principles predictions of spectra. Computational methods and the observed strength of turbulent fluctuations are discussed for a stack plume example. In another closeup, in Chapter 3, Peter Griffiths and co-workers, who have investigated the sensitivity of emission spectra to source temperature and who, as a result, conducted experiments on the behavior of linewidths with temperature, here combine these studies to provide quantitative information on this aspect of emission spectra.

In Chapter 4, Bob Spellicy and Merle Persky bring their experience in atmospheric transmission studies and field measurements to bear on some new emission data taken with the same interferometer system as operated by Herget to give further examples of the sensitivity of emission spectra observed through the atmosphere to changes in flame conditions.

Effective use of all of these emission techniques presupposes at least some knowledge of the temperature and concentration flowfields of the gaseous sources observed. Flowfield calculation methods are available and are progressing, but this progress is only achieved through comparisons with experiment. One of the most informative of these experimental measurements has been the emission–absorption technique reviewed by Chad Limbaugh in Chapter 5. This technique can provide both temperature and species concentration profiles. Not only could adaptations be used for continuous monitoring just as simple emission, but the full emission–absorption technique could also be used while a single-ended emission monitoring system was being set up, to characterize the source and establish a baseline.

Once the molecules identified in emission spectra from hot sources have mixed into the ambient atmosphere, their lower concentrations require sensitive absorption methods for detection. Although the high resolution spectroscopy discussed in Chapters 1–5 is still applicable, this is an example of an area in which it may provide more information than is actually needed. One of the simplest and most successful techniques, which, although still high resolution, provides only the (concentration) information desired is the gas-filter correlation analyzer. Dan Sebacher, whose recent applications of this method include detection of HCl in Space Shuttle launch clouds and methane emission from swamps, provides, in Chapter 6, a review of applications of the technique to date and the present state of the art in sensitivity to trace species and gives, in principle and example, the information needed to calculate sensitivity to yet other molecules.

Each chapter necessarily presents only a selection from the material needed for a thorough working knowledge of its subject. Therefore, particular care has been taken to include enough references to lead interested readers to the appropriate literature. The book is intended to be at the level at which the practicing engineer (or graduate student) with some physics background will be comfortable with most of it, consulting at times references to both more basic and more advanced literature. The book should alert readers to the existence of techniques that provide an abundance of information on questions of practical interest and that are becoming increasingly easy to apply, and stimulate some of them to look further into the methods it describes.

Joda Wormhoudt

Contents

Contributors

JOHN A. CONANT, Advanced Systems Group, Aerodyne Research, Inc., Billerica, Massachusetts

PETER R. GRIFFITHS, Department of Chemistry, University of California, Riverside, Riverside, California

LARRY A. GROSS, Department of Chemistry, University of California, Riverside, Riverside, California

WILLIAM F. HERGET, Nicolet Analytical Instruments, Nicolet Instrument Corporation, Madison, Wisconsin

CHARLES C. LIMBAUGH, Technology Department, Sverdrup Technology, Inc., AEDC Division, Arnold Air Force Station, Tennessee

UDAY MAHAGAOKAR*, Department of Chemical Engineering, University of Houston–University Park, Houston, Texas

MERLE J. PERSKY†, OptiMetrics, Inc., Bedford, Massachusetts

H. WILLIAM PRENGLE, JR., Department of Chemical Engineering, University of Houston–University Park, Houston, Texas

DANIEL I. SEBACHER, Atmospheric Sciences Division, NASA Langley Research Center, Hampton, Virginia

---

Current affiliations:
*Shell Development Company, Houston, Texas
†Lincoln Laboratory, Massachusetts Institute of Technology, Lexington, Massachusetts

SHUN-KWOK TSE*, Department of Chemical Engineering, University of Houston–University Park, Houston, Texas

ROBERT L. SPELLICY, Southwest Operations, OptiMetrics, Inc., Las Cruces, New Mexico

JEFFREY N.-P. SUN, Analect Instruments Division of Laser Precision Corp., Irvine, California

JODA WORMHOUDT, Research Group, Aerodyne Research, Inc., Billerica, Massachusetts

---

**Current affiliation*: Dupont Company, LaPorte, Texas

List of Symbols and Units

(The second column gives the equation defining or first using the quantity)

A	(3.3)	Absorbance (optical depth).
A_g	(2.1)	Area of target observed by telescope, m^2.
A_m	(2.1)	Area of receiving mirror, m^2.
B	(3.9),(4.10)	Rotational constant, cm^{-1}.
B	(2.12)	Background radiation factors.
C_1,C_2	(2.15),(5.76)	Planck blackbody function constants, 3.741×10^{-9} W, 1.4389 cm K.
C_2		
C_p	(2.26)	Heat capacity at constant pressure, J K^{-1} $mole^{-1}$
D	(3.9)	Centrifugal distortion constant, cm^{-1}.
D_j	(2.27)	Stack or jet diameter, m.
E	(3.8)	Level energy, cm^{-1}.
E	(2.4)	Electric field amplitude of wave at zero displacement, V/m.
F	(3.11)	Herman–Wallis factor.
G	(2.10)	Instrument gain.
H	(2.4)	Irradiance, W cm^{-2}.
J	(3.6)	Rotational quantum number.
J	(5.74)	Source function, N^0S/d.
$\bar{K}$	Sec. 4.4.4	Average absorption coefficient band model parameter, S/d, m^{-1}.
$\bar{K}_0$	(5.32)	Temperature-dependent average absorptivity band model parameter, S_0/d, m^{-1} atm^{-1}.
$\bar{K}^0$	(5.33)	Average absorptivity band model parameter at STP, S^0/d, m^{-1} atm^{-1}.
K	(3.40)	Voigt function.

K'	(3.40)	Dimensional constant of Voigt absorption coefficient, m^{-1}.
L	(5.42)	Total path length through emitting gas, m.
L_a	(2.1)	Distance from target to collector, m.
L_c	(2.2)	Reference length, m.
L_e	(2.1)	Equivalent path length, m.
L_t	(2.21)	Length of telescope pipe, m.
M	(2.30)	Mach number, u_0/u_s; u_s is the velocity of sound at 298 K and 1 atm = 332 m/s.
N	(3.37), (5.15)	Exponent for linewidth temperature dependence.
N_J	(3.81)	Number of molecules in rotational level J.
N	(2.1),(3.2), (4.9),(5.1)	Radiance, W m^{-2} sr^{-1}.
N^0	(2.15),(4.14), (5.1)	Planck's blackbody radiance, W/m^{-2} sr^{-1}.
N_{Ec}	(2.33)	Eckert number.
N_{Pr}	(2.31)	Prandtl number.
N_T	(2.16)	Normalized temperature signal.
P	(2.1)	Total radiant power (energy flux), W.
$P(\bar{\nu})$	(3.17),(4.3), (6.3)	Spectral radiant power, W/cm^{-1}.
$\mathscr{P}$	(4.18),(5.28)	Line strength distribution function.
Q	(3.8)	Partition function.
R	Sec. 3.5.3	Intermolecular distance, m.
R	(3.21),(6.1)	Resolution.
R	(2.10),(4.3)	Responsivity, V/W.
S	(3.7),(5.17)	Integrated absorption, line strength, cm^{-2}.
S^0	(3.7),(5.17)	Temperature-independent band strength at STP, cm^{-2} atm^{-1}.
S	(5.43)	Path-averaged band model strength coefficient, cm^{-2}.
S	(2.10)	Signal intensity, V.
S_T	(2.10)	Signal from target, V.
S_m	(2.12)	Peak-to-peak maximum interferogram signal, V.
T	(2.1),(3.8), (4.10),(5.1)	Temperature, K.
T	(2.29)	Dimensionless temperature.
U_R	(2.27)	Dimensionless radial velocity, u_r/u_0.
U_z	(2.27)	Dimensionless axial velocity, u_x/u_0.
V	(6.8)	Signal voltage, V.
V	Sec. 3.5.3	Intermolecular potential energy, J.
W	(4.15),(5.7)	Equivalent width, cm^{-1}.
$\overline{W}$	(4.18),(5.28)	Average equivalent width, cm^{-1}.

X	(2.19)	Signal ratio, pollutants to $CO_2 + H_2O$.
Z	(2.27)	Dimensionless axial coordinate distance, z/r_j.
Z_{cv}	(2.30)	Dimensionless core length for velocity.
Z_{ct}	(2.31)	Dimensionless core length for temperature.
a	(3.42),(5.37)	Voigt profile parameter.
c	(2.3),(3.9),(4.10)	Velocity of light, m/sec.
d	(4.15),(5.24)	Line spacing, cm^{-1}.
f	(5.22)	Curve of growth.
g	(3.17)	Instrument line shape function.
h	(3.9),(4.10)	Planck's constant, W sec^2.
k	(3.8),(4.10)	Boltzmann's constant, J/K.
k_T	(2.26)	Eddy thermal conductivity, J K^{-1} $mole^{-1}$. m^2/sec.
l	(3.4),(5.1)	Path length, m.
m	(3.7),(4.9)	Line index.
m	(3.33)	Molecular weight, kg.
n	(3.37)	Exponent in power-law intermolecular potential.
p	(3.34),(4.17)	Pressure, atm.
r	(2.2),(5.72)	Radial coordinate, m.
r_p	(2.2),(5.72)	Radius of plume, m.
r_j	(2.27)	Stack or jet radius, m.
r_m	(2.21)	Telescope primary mirror radius, m.
s	(4.14),(5.1)	Line of sight coordinate, m.
t	(2.3)	Half-period of interferometer moving mirror, sec.
u	(2.24)	Velocity, m/sec.
u_r	(2.24)	Radial velocity, m/sec.
u_z	(2.24)	Axial velocity, m/sec.
v	(3.6)	Vibrational quantum number.
w	(5.74)	Transmittance derivative function.
x	(5.18)	Optical path length.
y	(5.51)	Lindquist–Simmons function.
z	(2.24),(5.72)	Axial coordinate of plume or jet, m.
α	(5.7),(6.4)	Absorptance.
β	Sec. 4.4.4	Band model parameter.
γ	(3.13),(6.1)	Spectral line half-width at half maximum, cm^{-1}.
γ_c	(3.34),(5.14)	Collision-broadened linewidth, cm^{-1}.
γ_0	(3.34)	Temperature-dependent pressure broadening coefficient, cm^{-1} atm^{-1}.

γ^0	(3.44),(5.15)	Temperature-independent pressure-broadened linewidth at STP, $cm^{-1}\ atm^{-1}$.
γ_D	(3.33),(5.34)	Doppler linewidth, cm^{-1}.
$\bar{\gamma}$	Sec. 4.4.4	Frequency-averaged band model linewidth parameter, cm^{-1}.
Γ	(5.43)	Path-averaged band model linewidth parameter, cm^{-1}.
δ	(2.4)	Optical retardation, path difference between waves in interferometer, m.
δ_0	(2.3)	Maximum optical retardation, m.
ε	(2.9)	Total emissivity.
$\varepsilon_{\bar{\nu}}$	(3.1)	Spectral emissivity.
ε_v	(2.25)	Eddy kinematic viscosity, m^2/sec.
ε_0	(2.4)	Electric permittivity of free space, 8.85×10^{-12} farad/m.
η	(2.17)	Dimensionless temperature function.
θ_v	Table 2.2	Vertical angle measured on tripod gearhead.
θ_h	Table 2.2	Horizontal angle measured on tripod gearhead.
κ	(2.1),(3.1) (4.17),(5.1)	Absorption coefficient, m^{-1}.
κ_0	(4.17)	Absorptivity, $m^{-1}\ atm^{-1}$.
ν	(2.3)	Electromagnetic frequency, sec^{-1}.
ν_d	(2.3)	Detector frequency, sec^{-1}.
$\bar{\nu}$	(2.1),(3.1), (4.9),(5.1), (6.1)	Wavenumber, cm^{-1}.
$\Delta\bar{\nu}$	(2.20),(4.14), (5.4),(6.1)	Frequency interval or bandpass, cm^{-1}.
$\bar{\nu}_H, \bar{\nu}_L$	(2.9)	High and low frequency limits, cm^{-1}.
ξ	(3.43),(5.38)	Voigt profile parameter.
ρ	(2.29)	Density, kg/m^3.
P	(2.27)	Dimensionless density.
$\boldsymbol{\rho}$	(3.21)	Resolution parameter.
σ	(2.2)	Optical shape factor.
τ	(3.1),(4.13), (5.2),(6.3)	Transmittance.
$\tau(\bar{\nu})$	(2.9),(3.1)	Spectral transmittance.
$\bar{\tau}$	(2.14),(5.6)	Average transmittance.
χ	(4.17)	Mole fraction.

NOTE. The symbols for radiometric quantities used here are not those of the SI system, but are those taken from the American Standards Association (now ANSI, Standard Z58.1.1-1953) by the Working Group on Infrared Backgrounds [1]. This set of symbols has been common in defense-related work and is that used in the NASA Handbook [2]. This means, for instance, that instead of the SI symbols for flux, radiance, and irradiance of Φ, L, and E, we use P, N, and H. Reference works which do use the SI symbols include those by Wolfe and Zissis [3] and Nicodemus [4].

REFERENCES

[1] *Infrared Target and Background Radiometric Measurements: Concepts, Units, and Techniques*, Report Number 2389-64-T, Institute of Science and Technology, The University of Michigan, Ann Arbor, Michigan, January 1962, available DTIC, AD 275810.

[2] Ludwig, C. B., Malkmus, W., Reardon, J. E., and Thompson, J. A. L., *Handbook of Infrared Radiation from Combustion Gases*, NASA SP-3080, 1973.

[3] Wolfe, W. L., and Zissis, G. J., The Infrared Handbook, prepared by Environmental Research Institute of Michigan, Ann Arbor, Michigan, for the Office of Naval Research, Department of the Navy, Washington, D.C., 1978.

[4] Nicodemus, F. E. (ed.), Self-Study Manual on Optical Radiation Measurements, NBS Technical Note 910, March 1976, and Proc. SPIE *344*, 127 (1982).

INFRARED METHODS FOR GASEOUS MEASUREMENTS

1

High Resolution Infrared Emission from Gaseous Sources

JODA WORMHOUDT and JOHN A. CONANT *Aerodyne Research, Inc., Billerica, Massachusetts*

WILLIAM F. HERGET *Nicolet Instrument Corporation, Madison, Wisconsin*

1 INTRODUCTION

Environmental concerns about the effects of gaseous emissions from stationary combustion sources span the range of scales from global (CO_2 green-house effects and N_2O-induced stratospheric ozone depletion), through continental (SO_2- and NO_x-induced acid rain), to regional (NO_x and unburned hydrocarbon-induced photochemical smog) and local (CO health effects, HCl and H_2SO_4 acid corrosion, and CO, PAH, and chlorinated hydrocarbon health effects). A clear need exists for versatile and sensitive monitoring devices to quantify gaseous emission levels in industrial smoke stack plumes and from waste gas flaring operations. This need will be enhanced by the growing trend to dispose of toxic chemical and industrial wastes by incineration rather than dump or settling pond disposal. Furthermore, the drive to maximize energy efficiency within the control constaints for various combustion-produced pollutants requires that combustor stoichiometries and temperatures be precisely monitored and controlled.

High spectral resolution remote sensing of infrared radiation from combustors, stack plumes, and external flares offers attractive possibilities for the monitoring of both pollutant emission levels and combustion parameters such as temperature and stoichiometry. The possibility of extracting accurate and precise information about relatively trace gaseous species levels as well as combustor operating conditions from infrared emission spectra has been greatly enhanced by the recent development of sensitive high resolution techniques such as Fourier transform infrared (FT-IR) spectrometers and laser heterodyne radiometers.

Previous work on this topic includes FT-IR observations of stack plume emissions (Low and Clancy, 1967), theoretical analyses of the utility of such FT-IR emission spectra to quantify stack plume temperatures and species concentrations (Chan, Lin, and Low, 1973), FT-IR laboratory and field measurements of power plant pollutant concentrations (Prengle et al., 1973), observed and calculated spectra of smoke stack plume and sky radiation (Selby and Reed, 1978), laboratory detection of gaseous pollutants with a CO_2 laser heterodyne radiometer (Menzies and Shumate, 1974), and the use of a CO_2 laser heterodyne receiver to determine combustor gas and wall temperature (Stein, 1981). Recently, use of a double-beam interferometer to suppress background radiation has been shown to be a valuable technique in observations of stack plumes which showed emission spectra from CO_2, SO_2, and HCl (Shepherd et al., 1981). The most extensive catalog of pertinent measurements published to date is the work of Herget and Brasher (1979, 1980) and Herget (1982), who have obtained the infrared emission spectra of a variety of gaseous pollutant sources using the EPA's mobile FT-IR system.

The main focus of this chapter is the spectrally detailed modeling of the FT-IR emission spectra of a typical power plant plume and waste gas flare as measured by Herget and Brasher (1979). By varying model input parameters and comparing the output's fidelity to measured spectra, it is possible to assesss the sensitivity of remote, high resolution infrared emission measurements to source parameters such as temperature, molecular species concentrations, soot loading, and dimensional scale, as well as observational parameters such as intervening atmospheric path length and spectral resolution. Although the spectra presented here are quantitative, the emphasis is on qualitative results. They indicate the support needed from radiation transport modeling in designing automatic monitoring systems which can convert infrared spectral observations into useful information. For example, we will find that for species like H_2O and CO (present in both the plume and the atmosphere) molecular emission from band wing regions is relatively immune to atmospheric absorption and is especially sensitive to source temperature. Concentration measurements of species like carbon monoxide, which can be considered a pollutant or an indicator of combustion efficiency, are often complicated by nonlinearity in the scaling of spectral feature intensity with source concentration or size. However, compensation can be made for these nonlinearities through modeling. Other examples will show that heavy particulate loadings in the observed hot gas source can wash out the molecular emission features. Modeling can indicate when this can be a problem and show which spectral regions are least affected.

Our ability to accurately model the detailed infrared emission spectra of combustion sources draws on extensive studies of the exhaust plume heat transfer properties and spectral signatures of space vehicles, military rockets, and jet aircraft. This background is reviewed in Section 2.

2 TECHNOLOGY BASE

Developments in several areas of aerospace technology have led to major advances in the ability to calculate infrared radiation emission from hot gases. This brief review of this work outlines the technology base which can support applications of infrared emission techniques to environmental measurements and process control.

An early requirement of the space program was estimation of heat loads on the rocket due to heat transfer from the exhaust plume by radiation (Rochelle, 1967). Rocket exhausts are typically fuel-rich and thus contain fuel molecules such as H_2 and CO. Upon mixing with the atmosphere, afterburning produces molecules such as H_2O and CO_2, which in certain spectral bands are very efficient infrared radiators. Rocket plumes may contain no species that emit outside of these bands, so that it was clearly necessary to take into account the spectral structure in heat transfer calculations. The emission spectra of the CO_2 and H_2O molecules are composed of so many closely spaced lines that it is appropriate to treat them as continua, using molecular band models. Band models that were originally developed for heat transfer calculations continue to be standards in many applications; they are documented by Ludwig et al. (1973), in the NASA Radiation Handbook. [Progress in band modeling continues, as in the more recent CO_2 and H_2O band models by Bernstein (1980) and Young (1975).]

The NASA Handbook is also a standard source of radiation transport theory and nomenclature, as well as information on molecular line radiation and the optical properties of particles found in many rocket exhausts—carbon from liquid propellants and alumina from solid propellants. In general, the optical properties of these particles are less well known than the optical properties of gaseous species, and particles can scatter radiation as well as absorb and emit; therefore, this aspect of the radiation transport problem continues to receive special attention (see Jones, 1979; Watson and Lee, 1977; Huffaker, 1966; Reardon, 1970). Most heat transfer problems, however, can be adequately treated using a standard computer code also written for NASA (Huffaker, 1968).

The development of a second area of plume technology, that of radio transmission and radar reflection, resulted in computer codes which provided more detailed knowledge of plume flowfields. This occurred because both absorption and scattering of radio waves are influenced by the detailed chemical kinetics of plume afterburning through thermal and chemi-ionization processes. With these realistic flowfields, more accurate infrared emission spectra could also be calculated. These characteristic emission spectra (or infrared "signatures") were needed for the design of infrared missile detection devices (Gersh et al., 1980, referred to here as the JANNAF Handbook). A number of flowfield computer codes were developed. Similarly, the development of heat-seeking missiles led to a number of computer codes for the calculation of infrared radiation from aircraft engine exhaust plumes, where no afterburning occurs and the temperatures of the more dilute radiating species are determined through mixing with the atmosphere (Heck and Ferguson, 1971; Stone and Tate, 1975). However, the most widely used code for missile and aircraft plume signatures as well as radar studies became the Air Force Rocket Propulsion Laboratory Low Altitude Plume Program (LAPP) (Mikatarian, Kau, and Pergament, 1972), which calculated chemical kinetics and constant-pressure turbulent mixing for axisymmetric jets. Temperatures and species concentrations calculated with LAPP were input to variants of the NASA radiation code or to other codes with differing methods of approximating the geometry of radiation transport (Bernstein, Wormhoudt, and Conant, 1979; Pergament et al., 1979; Scanlon and Slack, 1975).

A substantial improvement over the original LAPP is represented by the BOAT code (Dash and Pergament, 1978), so named because of its development in a NASA program to study flow around rocket nozzle boat-tails. This code includes a number of the advances in turbulent mixing modeling which have also been incorporated into the parallel British effort in calculation of rocket exhaust flame properties (Jensen and Jones, 1981; Jensen et al., 1979). Both BOAT and LAPP have also been applied to the calculation of radiation from large stationary flames such as hydrogen burnoff flares (Fishburne and Pergament, 1978; Pergament and Fishburne, 1978).

Most recently, a tri-service program has resulted in a new generation of industry standard flowfield and radiation codes. The Standard Plume Flowfield (SPF) code (Dash and Pergament, 1980; Dash et al., 1980) contains BOAT, a calculation of the particle flow, and a method for including the shocks which result when high pressure exhausts expand into the atmosphere. The Standard Infrared Radiation Model (SIRRM) (Freeman et al., 1979; Ludwig et al., 1981) provides methods for radiation transport which include scattering by particles. (The Appendix gives information on obtaining these and other codes.) The improvements in this latest round of code development mostly address problems which are peculiar to rocket exhausts and are not found in the lower speed flows and less extreme environments of industrial flames and stack gases. An exception is the SHIP code (Pearce, 1980), a derivative of BOAT which treats the curved flows which can result when stack plumes are bent over by wind.

3 RADIATION PHYSICS

In this section we review some of the basic physics that influences the appearance of high resolution spectra. More detailed discussions are contained in the NASA and JANNAF handbooks. No equations will be used in this chapter, but each succeeding chapter will introduce more of the formalism. Introduction of radiation transfer formulae will begin in Chap. 2, while Chap. 3 includes, among other things, the molecular physics discussed below. Chapter 4 introduces band model concepts, while Chap. 5 presents a fuller exposition of them.

To begin, the effect of increasing temperature on gas emission is twofold. If the spectral emissivity of the gas were constant, then the emitted intensity of a volume of gas at a uniform temperature would simply increase according to the Planck blackbody function. However, molecules emit in bands whose individual lines originate in excited rotation–vibration states. In typical flames and plumes these states are in local thermodynamic equilibrium and have populations determined by the Boltzmann distribution function. With increasing temperature, more states are populated and the bands spread in frequency as more lines emit. As the population is distributed over more states, the intensity of individual lines near the band center drops, while the intensity of lines in the wings increases. For species present in both the plume and the atmosphere, the cold atmosphere will absorb only near band centers and will emit only weakly. Clearly the spectral regions of interest are those in which molecules in the gas source emit but in which atmospheric absorption is small, that is, where emission is due to hot molecules occurring on the long- and short-wavelength sides of the central absorption due to cold molecules in the atmospheric transmission path. For molecules like CO_2 and H_2O, which are present in the atmosphere and whose emission spectra are composed of many overlapping lines, the features of interest are referred to as the red and blue spikes. For molecules like CO, which emit in individual lines rather than in unresolved bands, it is still the lines far from a band center which suffer the least atmospheric self-attenuation, although the effect is not as drastic as for CO_2.

Of course, both flames and plumes often contain particulate emitters whose emissivities, while not necessarily wavelength-independent, are at least slowly varying over the entire infrared region. Unlike molecules with well-defined optical properties, such particles can have wide ranges of size distribution and composition, so that although some data are available, these radiators are not well characterized. However, particle property investigations have yielded information on soot in flames (Dalzell and Sarofim, 1969; Hottel and Sarofim, 1967; Chippett and Gray, 1978; Howarth, Foster, and Thring, 1966; Siddall and McGrath, 1963) as well as similar carbon particles in rocket exhausts (Stull and Plass, 1960; Boynton, Ludwig, and Thomson, 1968). In smoke stack plumes the particle variety can be even larger, including not only soot but also particles such as water droplets, coal or wood ash, or cement dust (Tien, Doornink, and Rafferty, 1979; Palmer, 1976; Pilat and Ensor, 1971; Wall et al., 1981; Gladney, Small, and Gordon, 1976; Schulz, Engdahl, and Frankenberg, 1975).

The temperature effects outlined above are complicated in actual observations by the fact that real flames and plumes are not isothermal but have temperature profiles. This means that the temperature of the flame regions that actually are visible to an observer changes with the spectral variation of transmission through the gas. Understanding of these optical depth effects is crucial to interpretation of observed spectra.

The emission spectra presented here were calculated using a fast, sophisticated radiation transport code applicable to cylindrically symmetric geometries (Bernstein, Wormhoudt, and Conant, 1979). The scaling of spectra with various parameters will be investigated using simplified models of the radiating source, in which the temperature and concentration distributions will be represented by as few concentric isothermal cylinders as possible—generally one at a primary radiating temperature and one or two at lower temperatures to account for self-absorption by the source in the outer, cooler layers. This is acceptable not only because the technology of calculating flowfields for these low-speed, highly turbulent flows is somewhat less advanced than that for axisymmetric jets such as missile exhausts, but also because temperature gradients are in general expected to be less steep so that one or a few isothermal cylinders is a good approximation. Parametric calculations presented in Sec. 3 of Chap. 4 will also show emission spectra to be relatively insensitive to the detailed form of the roll-off to ambient temperature, at least for small sources. Finally, it will be seen here that the present calculations give sufficient agreement with the observations to be instructive, although it might be expected that calculations based on first-principles flowfields would do even better.

Although no detailed flowfield calculations were carried out and compared with data, we can be guided in parameter choices by earlier calculation and data analysis programs such as those carried out at Aeronautical Research Associates of Princeton (ARAP) on large flames (Pergament and Fishburne, 1978; Fishburne and Pergament, 1978). As an example of flowfield and radiation calculations for large methane flames, Fig. 1 shows temperature and CO_2 concentration profiles for such a flame, which was modeled as an axisymmetric jet. The profiles shown are for an axial position upstream of where the mixing and burning region has finally reached the jet axis, so that this region is marked by temperature and combustion product maxima. Still, the essentially Gaussian nature of such low-speed mixing profiles is already evident. An additional consideration in choosing single temperatures to represent the flame or plume core is that in spectral regions of high optical depth, only the outer portions of large scale sources will be seen. An example of this is given in Fig. 1b, in which the transmission through the flame at 2200 cm^{-1} (near the peak of the

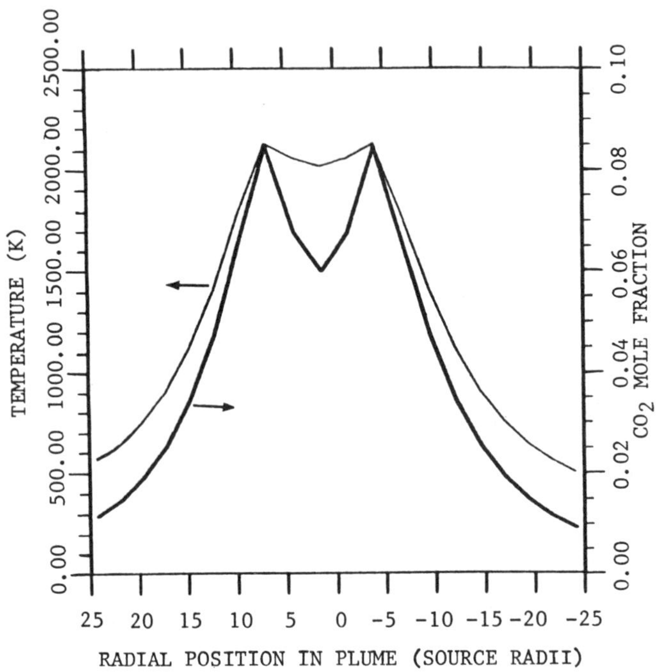

(a)

Fig. 1 Profiles and differential radiance for large flame calculation (Pergament and Fishburne, 1978; Fishburne and Pergament, 1978), for 90° line of sight at 200 source radii downstream, at 2200 cm^{-1} (near peak of CO_2 red spike). Source radius is 40 cm. (a) Temperature and CO_2 mole fraction profiles.

CO_2 4.3-μm red spike) is plotted, together with the differential contribution to radiance along this line of sight. It can be seen that the largest contribution to the signature at this spectral point comes from an area at around 1000 K, while the peak temperature region contributes little.

The fact that different temperature regions contribute differently to the spectra suggests that numerical methods can be used to invert emission spectra to yield temperature and concentration profiles, or to determine one when the other is specified by measurement or calculation. However, while very accurate inversions can be made for one variable, leaving both unspecified can lead to severe nonuniqueness problems. Both results have been demonstrated in calculational and experimental studies of a furnace sys-

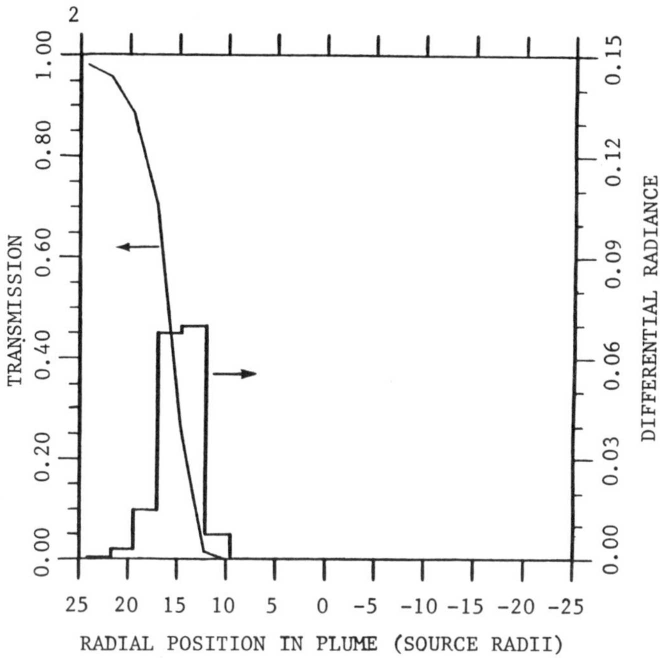

(b)

Fig. 1 (continued) (b) Transmission into flame, and differential radiance (contribution from each point along line of sight).

tem (Hommert and Viskanta, 1978). Chapter 5, whose main focus is inversion using multiple lines of sight in a single frequency band, will make the point that inversion using several spectral bands along one line of sight is only practical when coupled with a priori knowledge of the flowfield, either through observations of similar sources, correlations, or computer calculations. In two studies which investigated stack plume emission spectra to determine SO_2 concentrations, the temperature was either measured (Barnes, Herget, and Rollins, 1974) and assumed to be uniform across the plume or a single parameter in the temperature profile was deduced from the intensity of the blue spike of the CO_2 4.3-μm band (Tanabe, 1975). Once again, much of the progress in developing inversion methods has come from the desire for such information on rocket and exhaust plumes (Young, 1981). A review of these methods will be presented in Chap. 5.

At present, it appears that the most accurate information can be extracted from emission spectra by a combination of a priori prediction and inversion methods rather than by either one alone. Although they clearly deserve more study, this chapter will not treat inversion methods but will present complementary information which suggests spectral regions that are most promising for remote sensing.

4 OBSERVATIONS AND DETAILED MODEL CALCULATIONS

The remainder of the chapter will concern itself with parametric variations of radiation predictions for two sources: a refinery waste gas burnoff flare and a coal-fired power plant stack plume.

The observed spectra used to compare with model predictions were taken by one of us, W.F.H. (while employed at the Environmental Protection Agency), using a Nicolet Model 7199 Fourier transform interferometer spectrometer system configured to fit into an existing van (Herget and Brasher, 1980; Herget, 1979). The interferometer is matched to a Dall–Kirkham telescope (0.6-m diameter primary mirror) to collect infrared energy either from a remotely located light source (with identical telescope) or from warm gases exiting industrial stacks. A flat tracking mirror mounted on the side of the van is used to assist in sighting stack exits (or to obtain solar spectra). Measurements of gas pollutant concentrations have been made at a variety of pollutant sources (Herget, 1979 and 1982; Herget and Brasher, 1979 and 1980). The mobile interferometer system, which has the acronym ROSE (Remote Optical Sensing of Emissions), has proven to be a highly useful and versatile instrument for characterizing source emissions.

Two examples of the measured spectra, for a waste gas burnoff flare and for a stack plume from a coal-burning power plant, are given in Figs. 2 and 3. Flares, whereby waste gases are burned at the top of a stack, are used in many industries as a relatively inexpensive way of converting obnoxious pollutants into H_2O, CO_2, and (in some cases) less obnoxious pollutants such as HCl. The flare does present a potential regulatory problem, in that it is difficult if not impossible to determine the flare efficiency by conventional sampling techniques. Thus a series of measurements were made with the ROSE system on a flare burning vinyl chloride to assess the potential of this method.

For the spectra obtained at a coal-burning power plant (Fig. 3), the ROSE van was located approximately 800 m from the

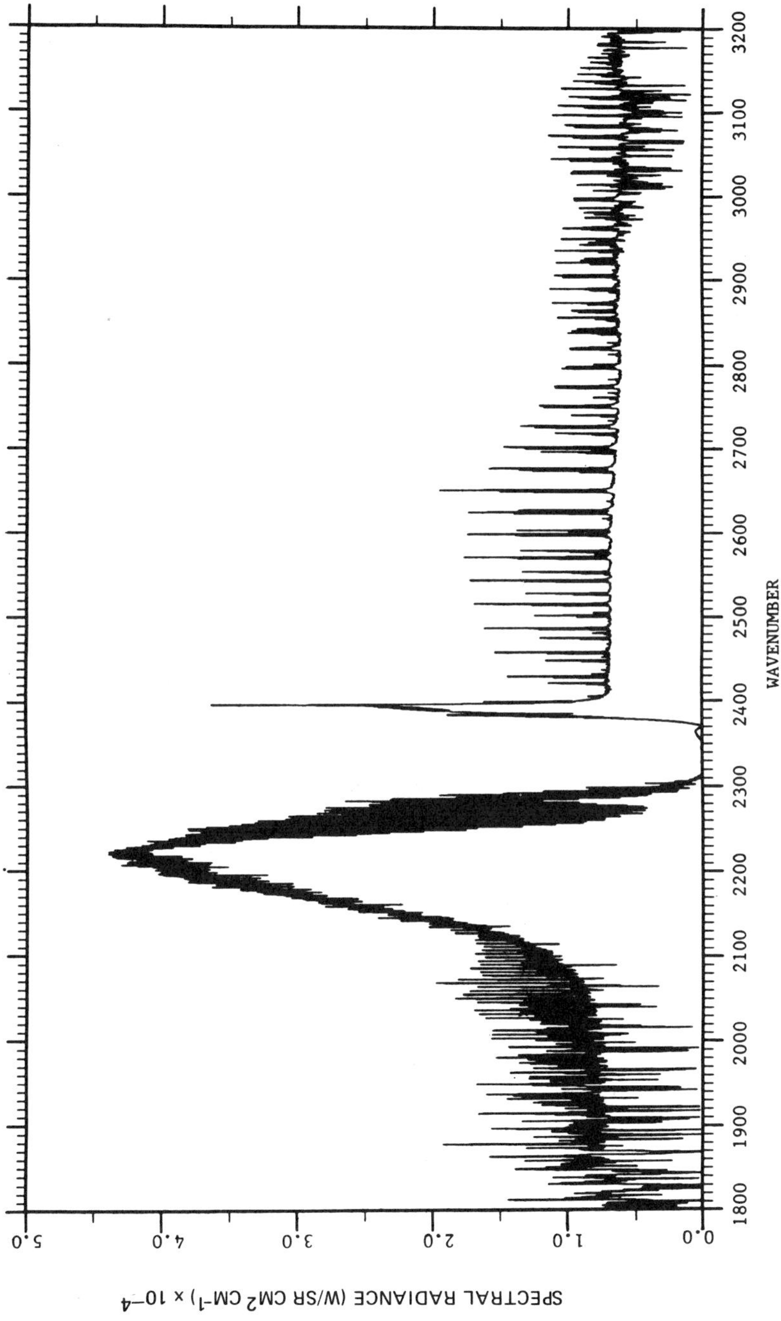

Fig. 2 EPA spectrum of waste gas burnoff flare.

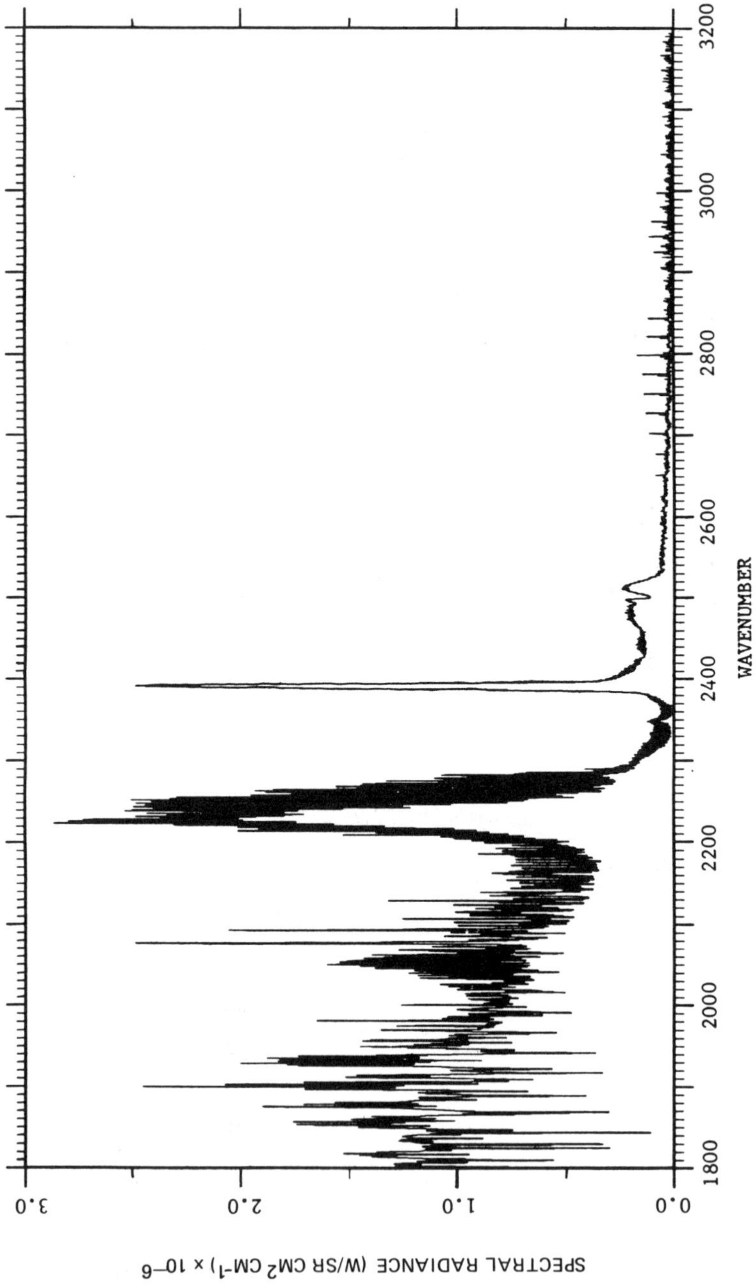

Fig. 3 EPA measured spectrum of coal fired power plant stack plume.

stack. The stack diameter was 6 m, the nominal effluent temperature was 400 K, and the visual opacity of the plume was approximately 10%.

The spectra obtained were calibrated in terms of spectral radiance (W/cm^2 sr cm^{-1}). They must be interpreted as average radiance within the finite instrument field of view (FOV). For the waste gas flare measurement the FOV subtended a circle of 0.94 m in diameter, almost twice as wide as the 0.5-m burnoff stack. Calculation of the flare radiant intensity was made assuming the flame was 0.5 m in diameter and 0.94 m long. The predicted intensities were then divided by the FOV area of 0.69 m^2 to convert to radiance for comparison with the measured spectrum. The calculated flame core was taken to have a temperature of 1800 K, with infrared emitting species mole fractions of 0.10 for CO_2, 0.20 for H_2O, 0.003 for CO, 0.003 for HCl, and 5×10^{-4} for soot. HCl appears in the observed flame because vinyl chloride is included in the fuel. The observed soot concentration varied substantially depending on the operating conditions of the burnoff flame, and indeed in adapting the above model to other fuels and combustion conditions, this is a parameter where large variation would be expected.

For the power plant stack plume measurement, the instrument FOV subtended a 4.1-m diameter circle, compared to a 6-m diameter stack. The information for the power plant stack plume was translated into a simple model in which the inner (400 K) cylinder had a diameter of 0.80 of that of the stack, with a length of 37.5 m, species mole fractions for CO_2 of 0.13, for H_2O of 0.06, and for HCl produced by combustion of chlorides in the coal, 8 ppm. Although CO lines were seen, their intensity is small compared to CO_2 and H_2O emission in the same region (not suprisingly, since power plants maintain combustion efficiency by minimizing CO emissions), and therefore no CO was included in the baseline model. However, CO content will be a parameter in the studies described below, as external burning sources might be expected to have appreciable CO in their smoke plumes, and in parametric variations, the nominal CO mole fraction will be 10 ppm. Although this concentration is typical, observed CO stack emissions vary over several orders of magnitude (Cato et al., 1974; Environmental Protection Agency, 1973; National Air Pollution Control Administration, 1970; Barrett, Miller, and Locklin, 1973). Two cooler, concentric outer cylinders were also included in the model inputs. Predicted spectral radiant intensities were subsequently converted to radiances by dividing by the 180 m^2 model plume area.

The calculated spectra are shown in Figs. 4 and 5. Overall agreement with observed spectral shapes and intensities is reasonable. Even better agreement could be achieved by fine tuning some parameters, as will be seen in the following discussion of spectral sensitivity to source and observation parameters. It should also be noted that for the power-plant plume the modeled FOV included the entire plume width, unlike the measurement. The outer edges of the plume are expected to be relatively dim due to the shorter plume depth and the cooler temperatures generally found there. The modeled average radiance should thus be somewhat low.

5 PARAMETRIC STUDIES

Now we turn to a study of the effect on predicted emission spectra of variation of several of the model parameters from their nominal values, first for the burnoff flare and then for the stack plume. The spectra presented in this section will be simplified from those in Figs. 4 and 5 in several ways. First, instead of the high resolution atmospheric transmission calculation, a band model will be used which represents only the average attenuation over 5 cm^{-1} or larger intervals (CO line absorption is treated exactly). Second, HCl will not be included, as the predicted behavior of isolated emission lines is represented well enough by CO, which is present in almost all combustion gases. Furthermore, the 4-μm SO_2 emission band seen in the stack plume spectrum, a very useful feature for pollution monitoring, will not be modeled here. These simplifications have the effect of focusing attention on changes observed in the major species emission features which will now be examined at high resolution.

The first parameter varied to determine its effect on emission intensities and relative spectral structure is the length of the atmospheric transmission path from source to observer. The observation of Fig. 2 was taken at the fairly short path length of 100 m. Figure 6 shows predictions for one shorter (50 m) and two longer (800 and 2000 m) paths.

In examining the trends shown in Fig. 6, it can be seen that the largest qualitative differences in spectral shape can be found near the center of the CO_2 4.3-μm band. Two transmission maxima are observed. The first, at slightly less than 4.4 μm, is due to the $^{13}CO_2$ band center and is observed at short ranges. The second, slightly beyond 4.5 μm, is due to the atmospheric N_2O band center and is observed in all three spectra. It can also be

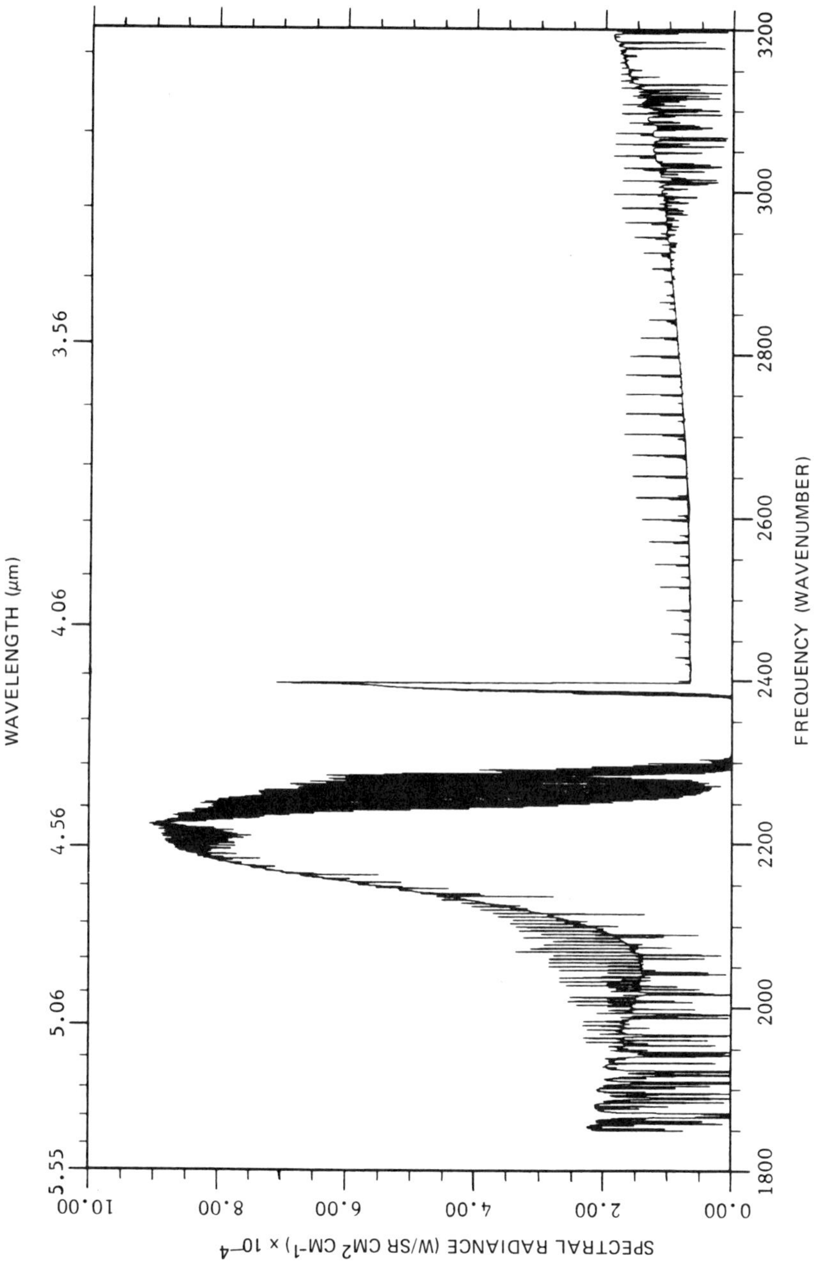

Fig. 4 High resolution calculated spectrum for gas burnoff flare.

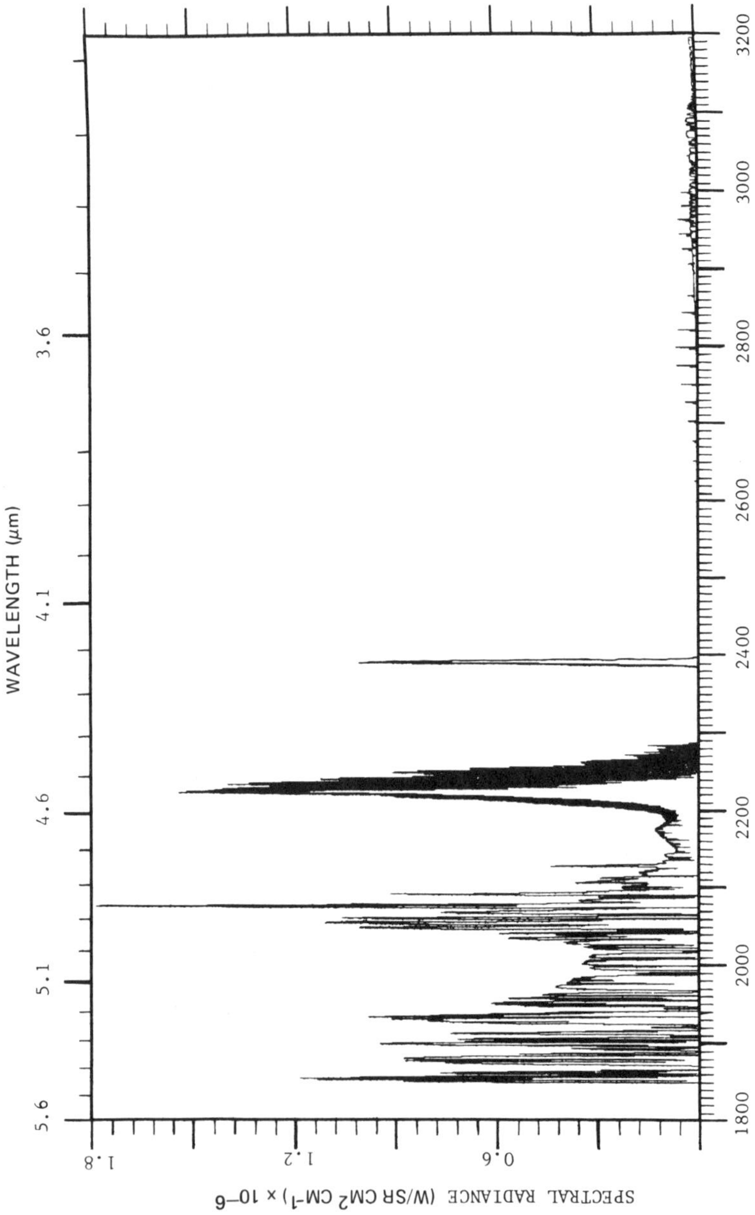

Fig. 5 High resolution calculated spectrum for coal fired power plant stack plume.

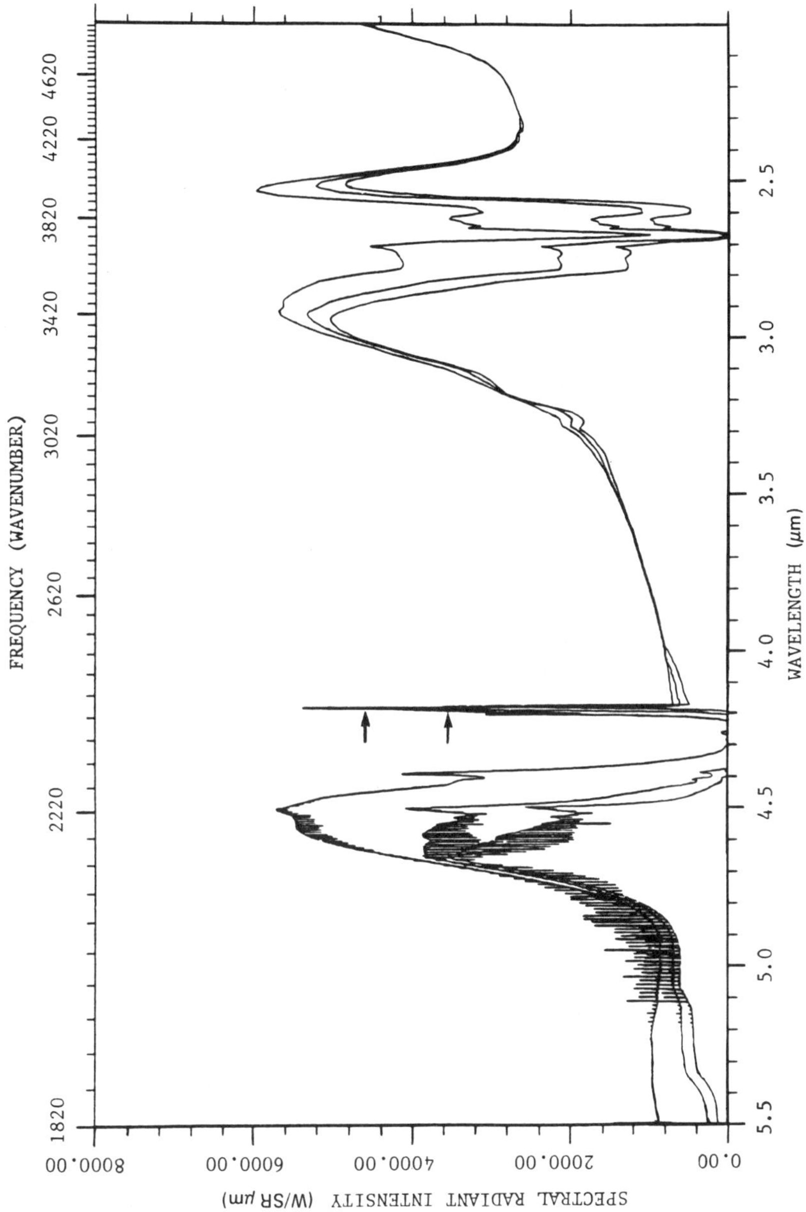

Fig. 6 Model flame spectra for 50-, 800-, and 2000-m atmospheric transmission paths. Arrows denote blue spike heights for two longer paths.

seen that at longer ranges the (low *J*) CO lines closer to the CO_2 band center reverse from emission to absorption. This can be seen more clearly in Fig. 7, which shows this part of the spectrum on an expanded scale, and is due to the effect of the naturally occurring 8×10^{-8} mixing ratio of CO in the atmosphere.

We continue by considering variations of CO concentration in the burnoff flame. Figure 8 shows the appearance of the CO emission region for models with two CO concentrations, corresponding to two possible degrees of combustion completeness. The lower value of 100 ppm, although predicting observable emission from some lines at 100-m range, would clearly be below the noise level of most real-time observations. At the other limit of 5 percent CO (in which CO_2 was reduced from 10 to 5 percent), the emission lines are clearly optically thick and influenced in intensity by the colder outer layers of the model flame, in that the ratio of line peak intensities to those of the nominal model with 0.3 percent CO is only about 5, rather than the ratio of more than 16 expected from the ratio of concentrations.

Next, we examine the variation of the spectral signature with core gas temperature. From Fig. 9 it can be seen that the major effects of increased flame temperature are an overall increase in intensity, a widening of the CO_2 red spike, and a decrease in the ratio of blue-to-red spike heights. More subtle spectral changes can be seen in the relationships between the $^{13}CO_2$ and N_2O band centers and their surrounding regions. Looking at Fig. 10, in which the CO emission region of Fig. 9 is reproduced on a larger scale, several changes with core temperature can be observed. Moving toward higher frequency, into the CO_2 red spike wing, which is especially sensitive to temperature, CO lines reverse into absorption sooner at higher temperatures. Toward lower frequency, the contrast between line and band emission stays roughly constant.

Using these spectral predictions, it is interesting to examine to what extent core temperatures could be immediately extracted. The point to be made is that because stronger lines are more highly absorbed in the cooler outer plume layers, the hot core contributes relatively less than it does to weaker lines. The ratios of line peak heights are, thus, not determined simply by line strengths and level populations (and hence temperatures). As an example, the two lines at the left of Fig. 10 are the 2-1 P(27) and 1-0 P(23) lines, which at a temperature of 1600 K (for an optically thin emitter) should have a peak ratio of 0.42. The ratio seen in Fig. 10 is actually 0.60, because the fundamental line is more strongly absorbed. Therefore, temperatures extracted from ra-

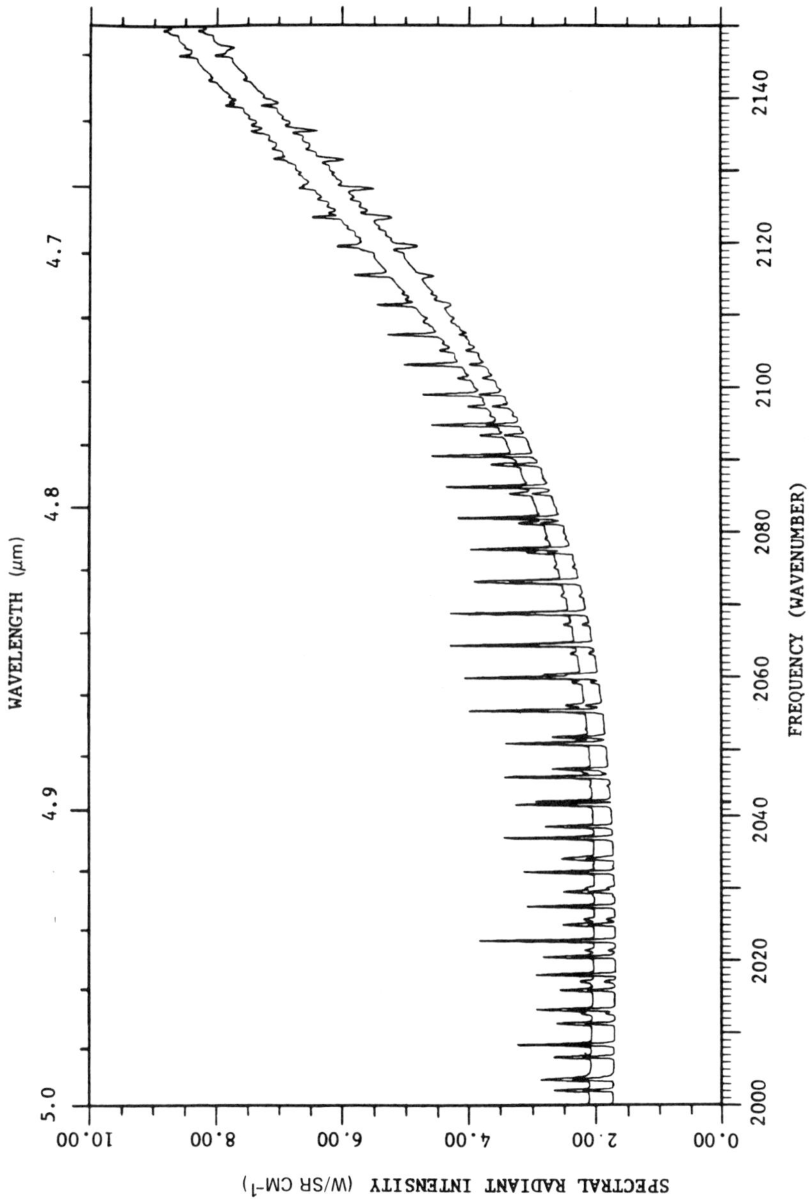

Fig. 7 CO emission region detail of model flame spectra for 50- and 800-m atmospheric transmission paths.

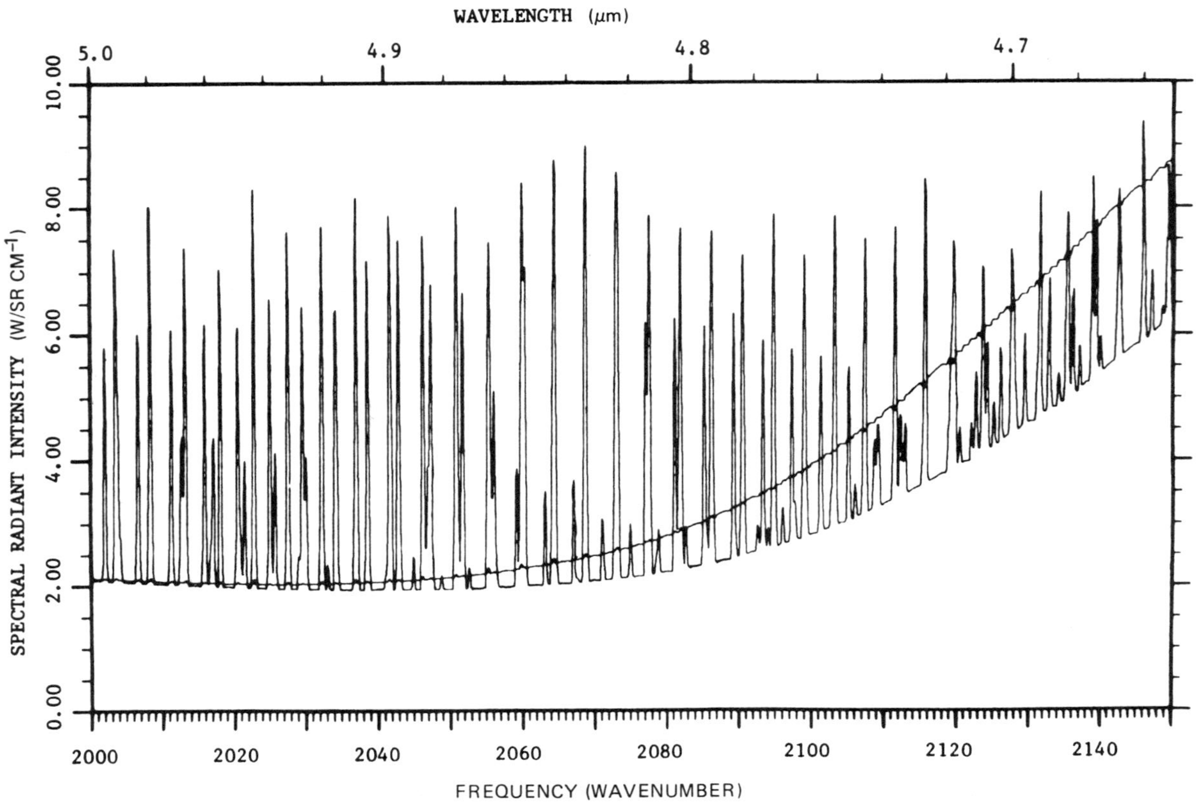

Fig. 8 Model flame spectra in the CO region for two limiting CO concentrations—100 ppm, and 5 percent (in which CO_2 is reduced from 10 to 5 percent).

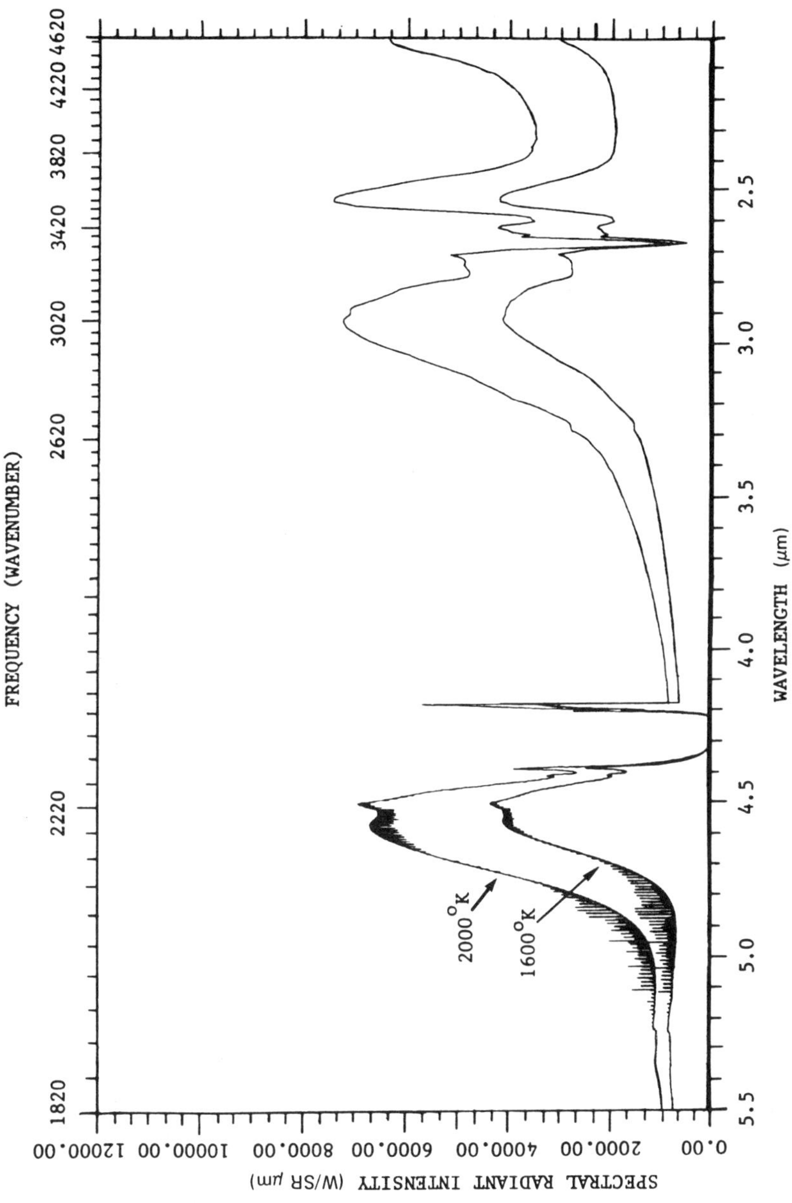

Fig. 9 Full model flame spectra for two alternative core temperatures—1600 K and 2000 K.

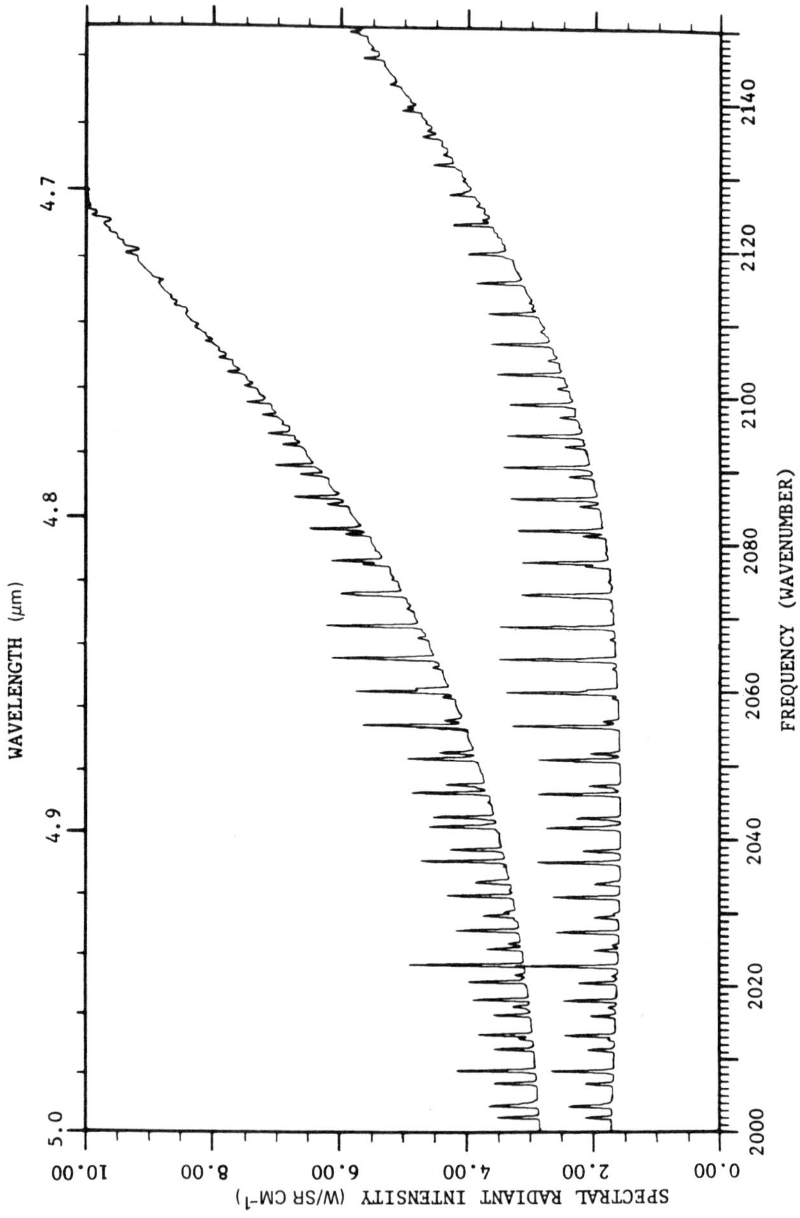

Fig. 10 Model flame spectra in CO emission region for core temperatures of 1600 and 2000 K.

tios of vibrational or rotational line pair peak intensities will be distorted by these optical depth effects, and good estimates of core temperatures from spectral data can only be obtained through modeling which takes into account realistic temperature profiles.

Another species concentration which can substantially affect high resolution spectral features, even though it has no such emission structure itself, is that of soot or other particulates produced by incomplete combustion. Figures 11 and 12 show the effects of increasing the nominal soot mole fraction of 5×10^{-4} by factors of 4 and 20. The smaller increase serves only to fill in areas of lower optical depth and is especially obvious at short wavelengths. The larger increase, however, is enough to add appreciable optical depth to the cooler outer layers of the flame, so that intensities are actually decreased in all spectral regions except the red wing of the CO band. In addition, the CO_2 blue spike and hot CO line emission features are almost obliterated.

The next parameter for study is the physical size of the flame. Decreasing all dimensions of the initially 50- by 125-cm flame by a factor of 2.5 (corresponding to a small flame) and even by 25 (to the size of a propane torch flame) results in essentially no qualitative change in the infrared spectra. Much more interesting in terms of qualitative spectral changes are those seen when flame sizes are scaled up beyond that of the relatively small nominal burnoff flare. Two examples of scaleup by factors of 4 and 10 are seen in Figs. 13 and 14. In the smaller of the two (200- by 600-cm projected area), some hot CO emission lines and a blue spike feature remain, even though the basic trend is to a blackbody spectrum with molecular absorption features. The larger example shows that very large flames, like those encountered in fires, might have no high resolution emission features.

Now we turn to a discussion of similar parametric variations on the nominal stack plume emission model. As before, we begin the spectral scaling studies with examples of the change in signature with atmospheric transmission path. Looking at Fig. 15, one can see some of the same trends observed earlier for flame spectra. For instance, the largest spectral differences occur near the center of the CO_2 4.3-μm band, and even the very small natural atmospheric CO concentration is enough to attenuate CO emission lines at long path lengths. Turning to the expanded scale spectrum of Fig. 16, one can see not only this effect on CO lines, but also some selective absorption of the low-temperature CO_2 emission features. The CO_2 bands in this region were calculated using 2-cm^{-1} resolution band models constructed from the AFGL line

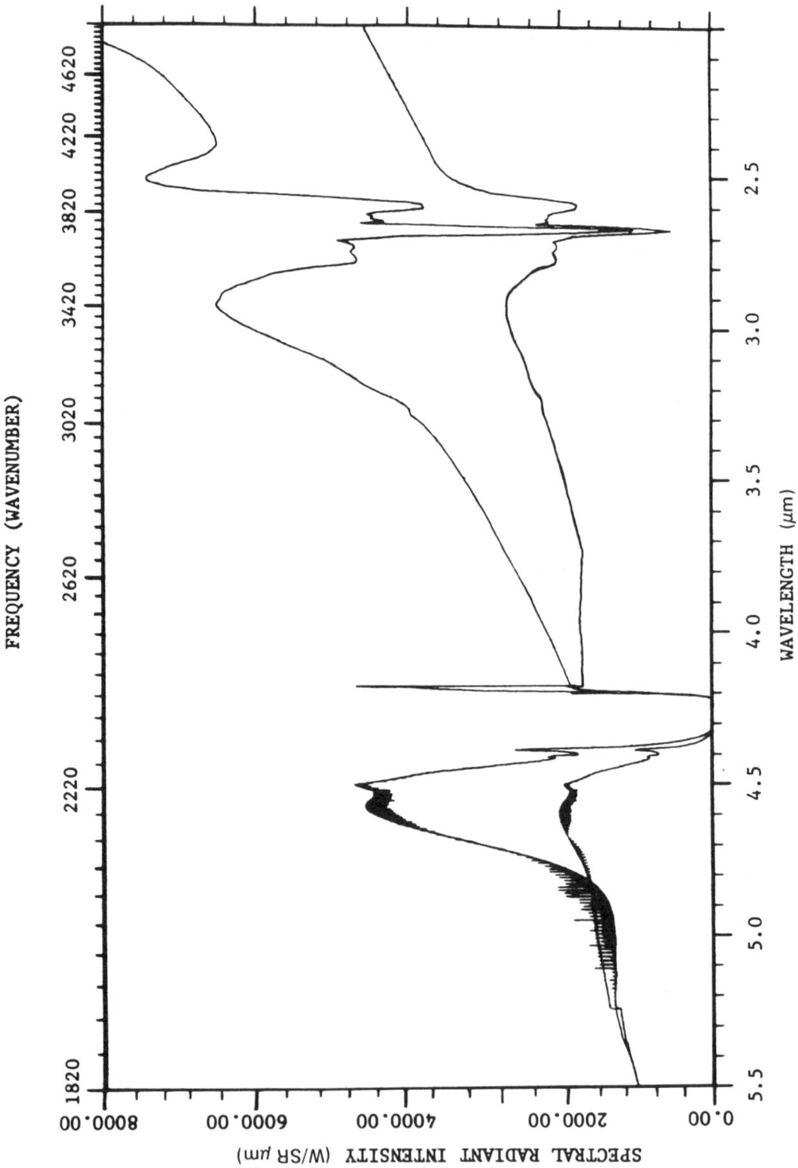

Fig. 11 Model flame spectra for two soot particle loadings above the nominal: 2×10^{-3} and 10^{-2} mole fractions for upper and lower curves, respectively.

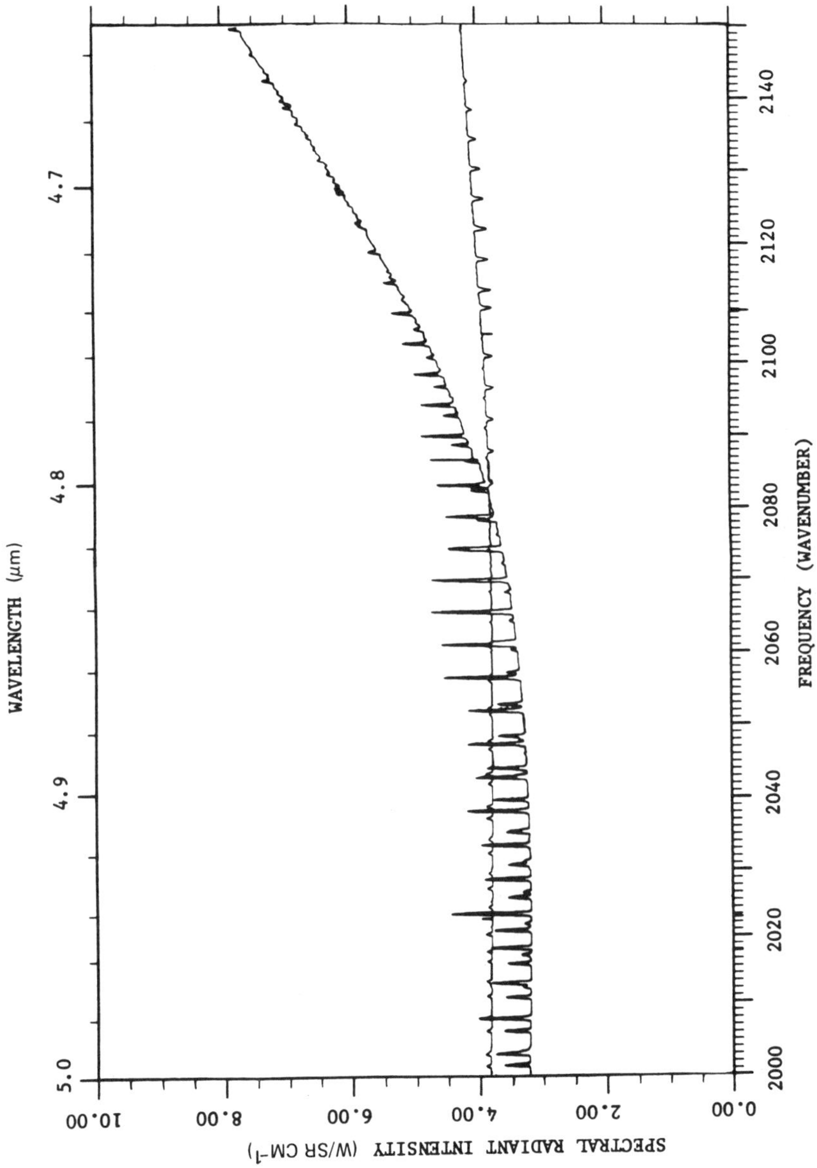

Fig. 12 Model flame spectra in CO emission region for 2×10^{-3} and 10^{-2} soot mole fractions.

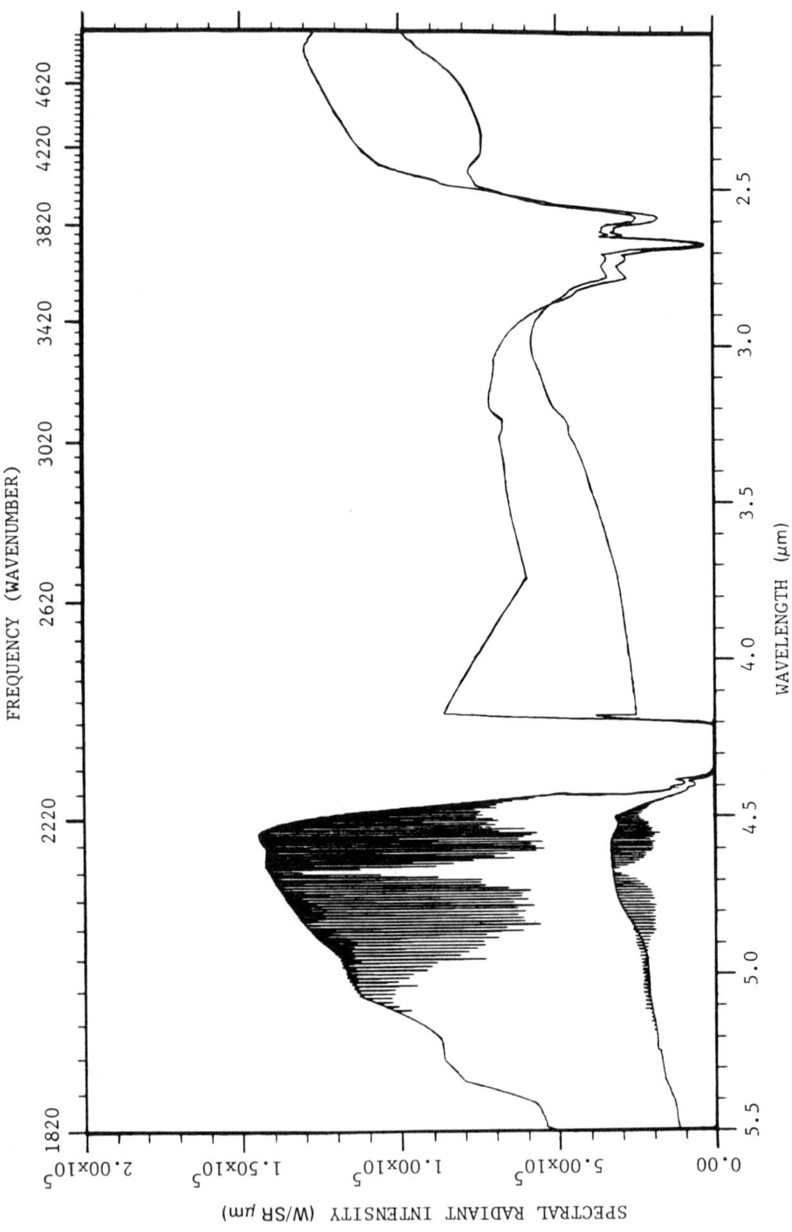

Fig. 13 Model flame spectra for two examples of large-scale flames—all dimensions increased by factors of 4 and 10.

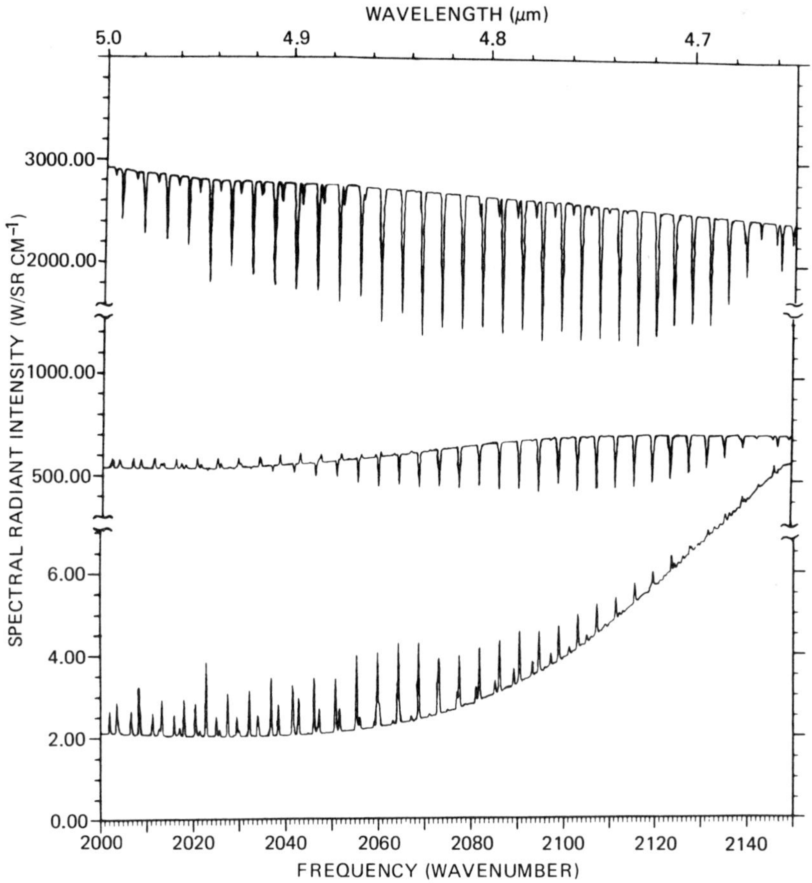

Fig. 14 Model flame spectra in CO region for nominal case (lowest curve) and two cases in which all dimensions increased by factors of 4 and 10.

parameter compilation (Rothman et al., 1983; see Appendix for availability), which accounts for their appearance as averages of groups of lines with occasional gaps. Three sets of broad bands and sharp spikes (Q branches) are superimposed, with only the largest, longest wavelength band clearly visible. Band identifications are indicated in Fig. 16. It can be seen that, as expected, the transition which is most highly attenuated terminates on a ground state, while the hot bands are less affected.

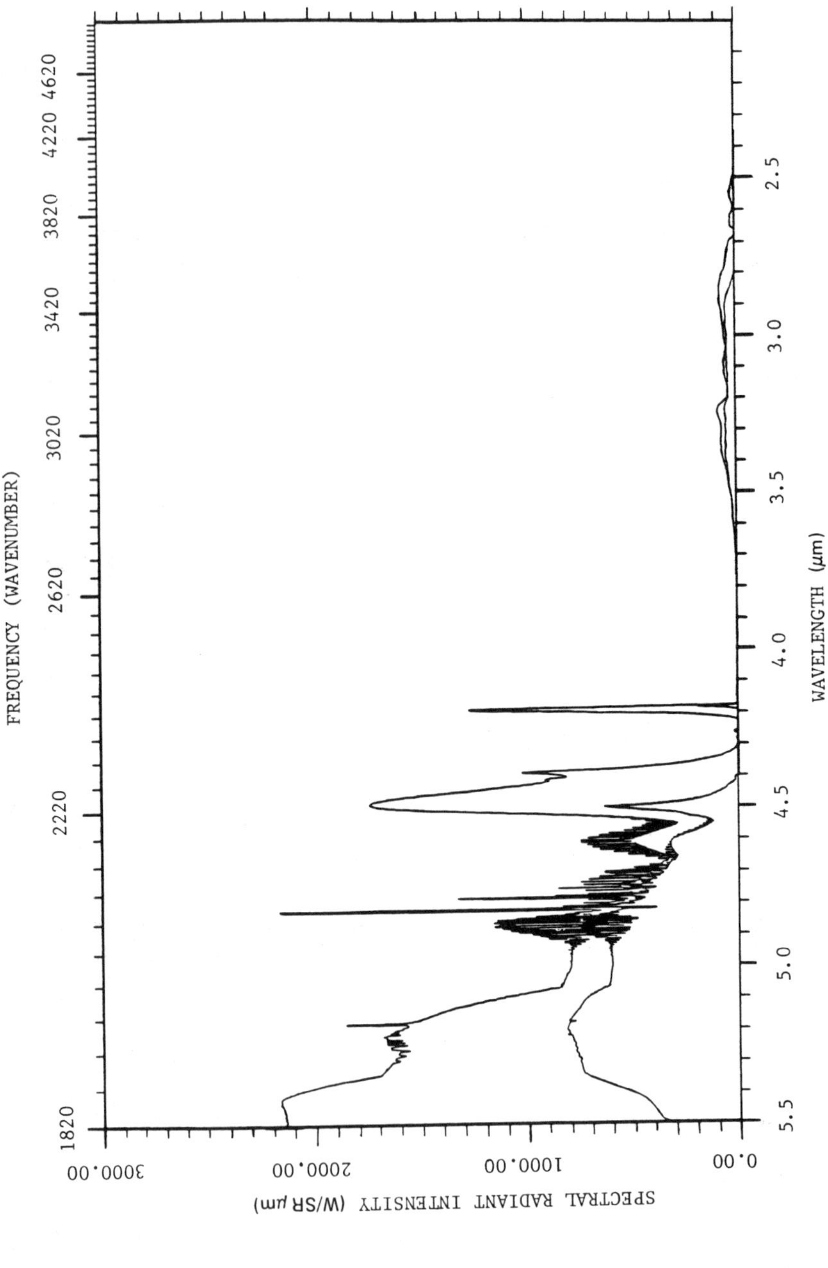

Fig. 15 Model stack plume spectra for atmospheric transmission path lengths of 50 and 2000 m.

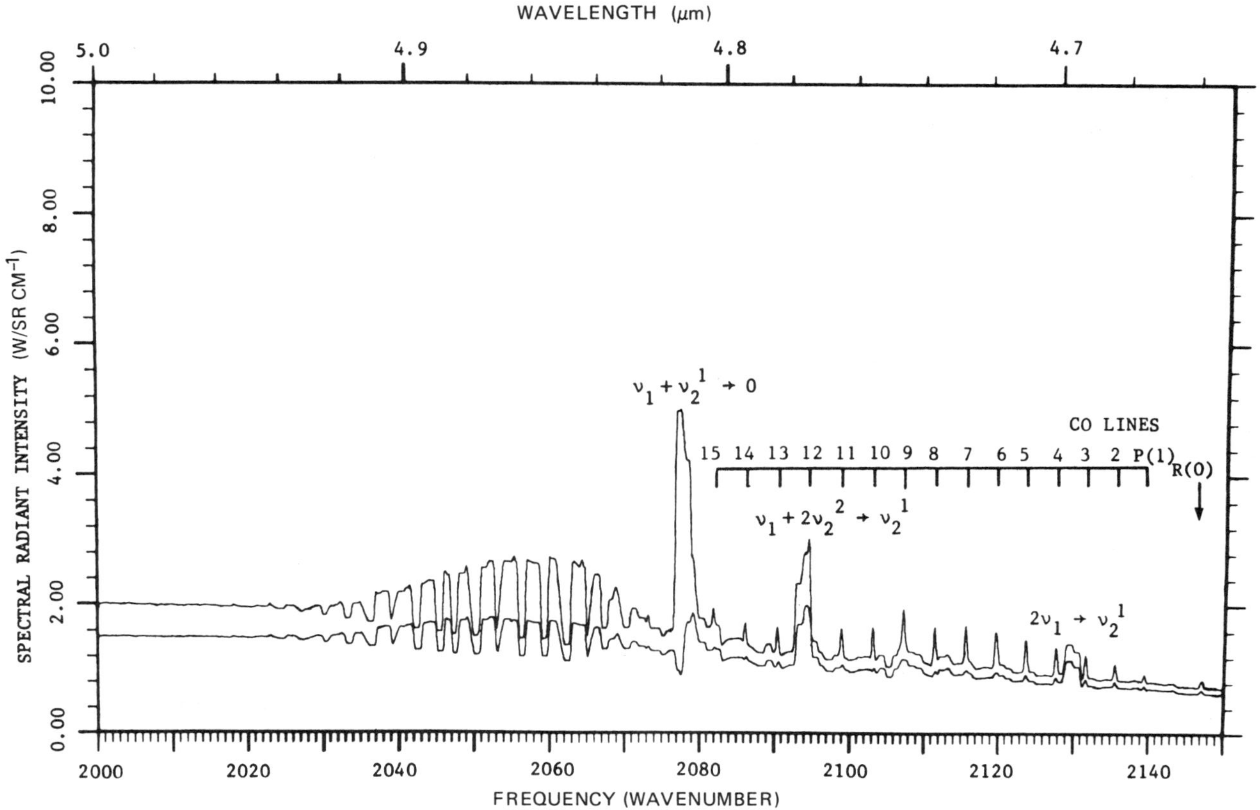

Fig. 16 Model stack plume spectra in CO and CO_2 emission region for atmospheric transmission path lengths of 50 and 2000 m.

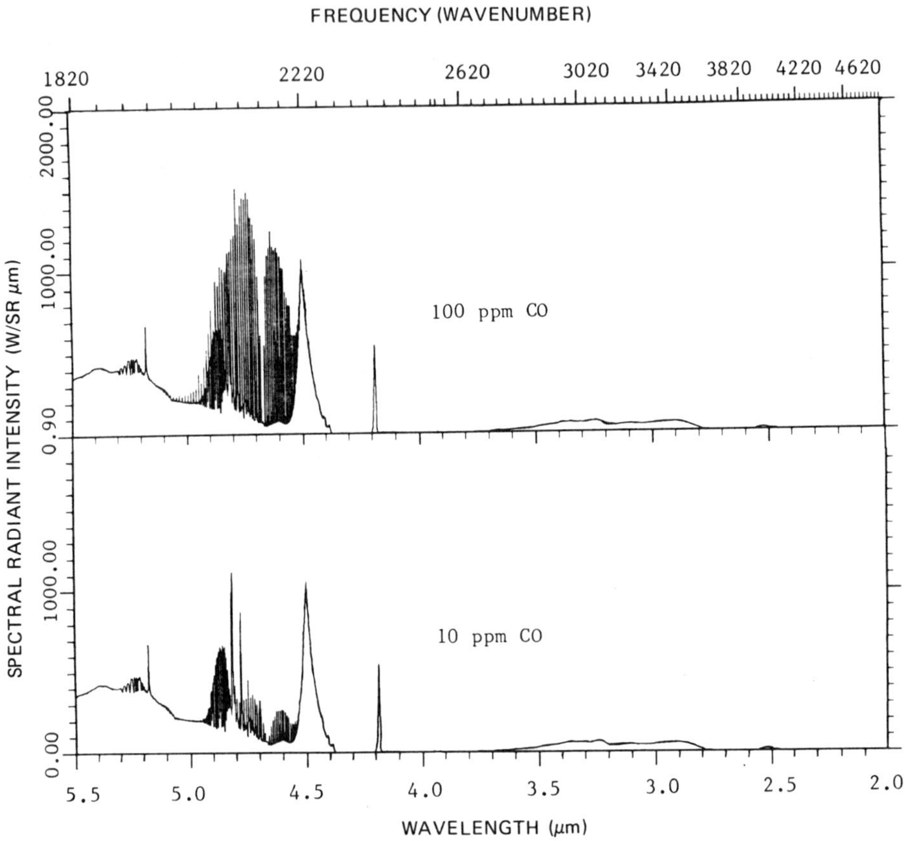

Fig. 17 Model stack plume spectra for CO concentrations of 10 (lower) and 100 ppm (upper).

Again, the CO concentration is of interest, since it can correspond to varying degrees of combustion completeness. From Fig. 17 it can be seen that 10 ppm is readily observable, and it turns out that even 1 ppm is easily detectable near the band center. The addition of more CO results in a less than linear increase in CO line emission intensity, as increasing plume optical depth leads to self-absorption by the cooler outer layers. This trend is already seen in Figs. 17 and 18, in which an increase from 10 to 100 ppm results in less than a full factor of 10 increase in line peak heights. Figures 19 and 20 show two even larger CO mole fractions, where now the fundamental lines have reached the opti-

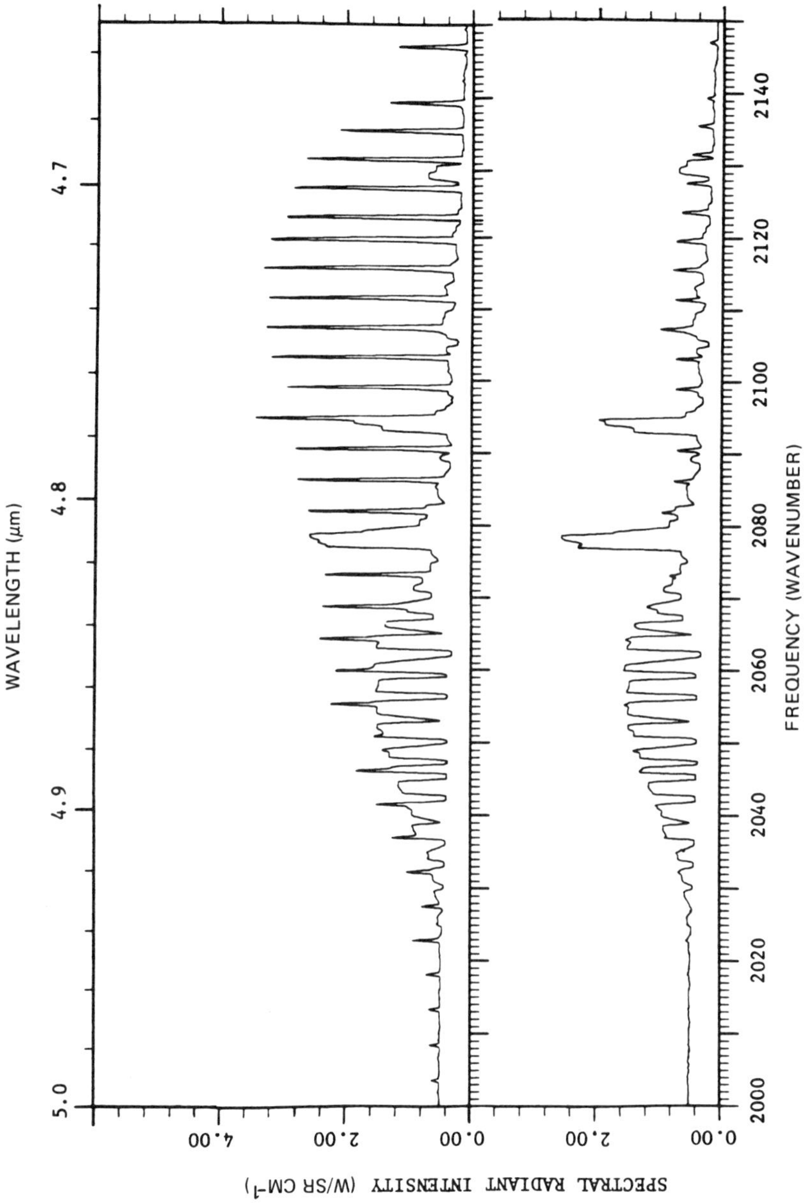

Fig. 18 Hot CO and CO_2 emission region of model stack plume spectra for 10-(lower) and 100-ppm CO.

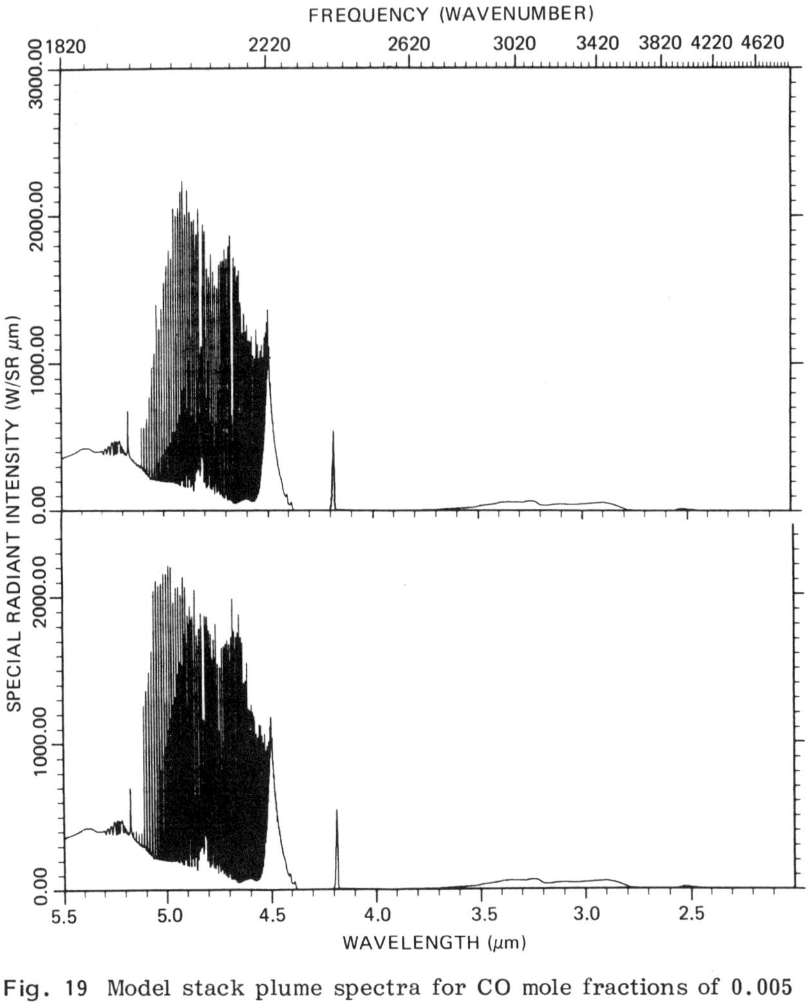

Fig. 19 Model stack plume spectra for CO mole fractions of 0.005 (upper) and 0.03 (lower).

cally thick limit, and the only changes in the spectrum result from continued growth in overtone lines. Once again, changes in intensity are by no means proportional to changes in concentration but rather are determined by the individual absorption coefficients of the various lines. For example, again for the two lines at the left of Fig. 20, the 2 → 1 vibrational transition peak height

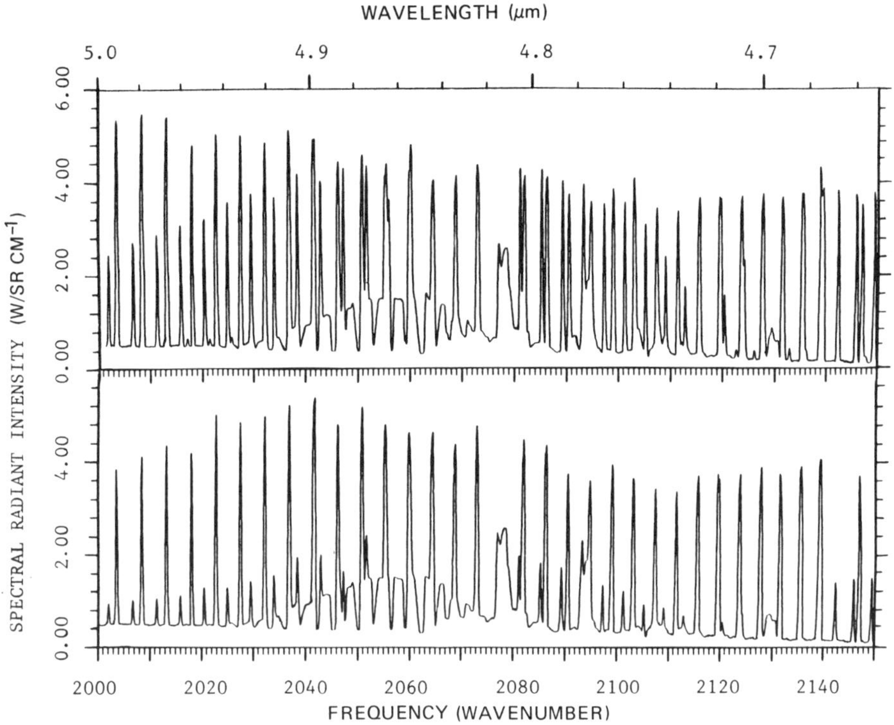

Fig. 20 Hot CO and CO_2 emission region of model stack plume spectra for 0.005 (lower) and 0.03 (upper) mole fractions CO.

increases by a factor of 4.3, while the neighboring 1 → 0 line increases by only 1.45, for a concentration increase of a factor of 6.

Next, we consider effects of differing plume core temperatures—clearly a parameter of importance, as smoke and exhaust plume temperatures can span the entire range between flame and ambient values. Figures 21 and 22 show examples of spectra appropriate to core temperatures of 350, 550, and 700 K (where the nominal value was 400 K). In Fig. 21 it can be seen that the sensitivity to temperature changes increases with decreasing wavelength, so that the long wavelength CO_2 and water bands are least affected, the CO_2 4.3-μm blue-to-red spike peak ratio changes from less than to greater than one, and a most dramatic increase is seen in the 2.7-μm water band. In both figures it can also be seen that with increasing plume optical depth at higher

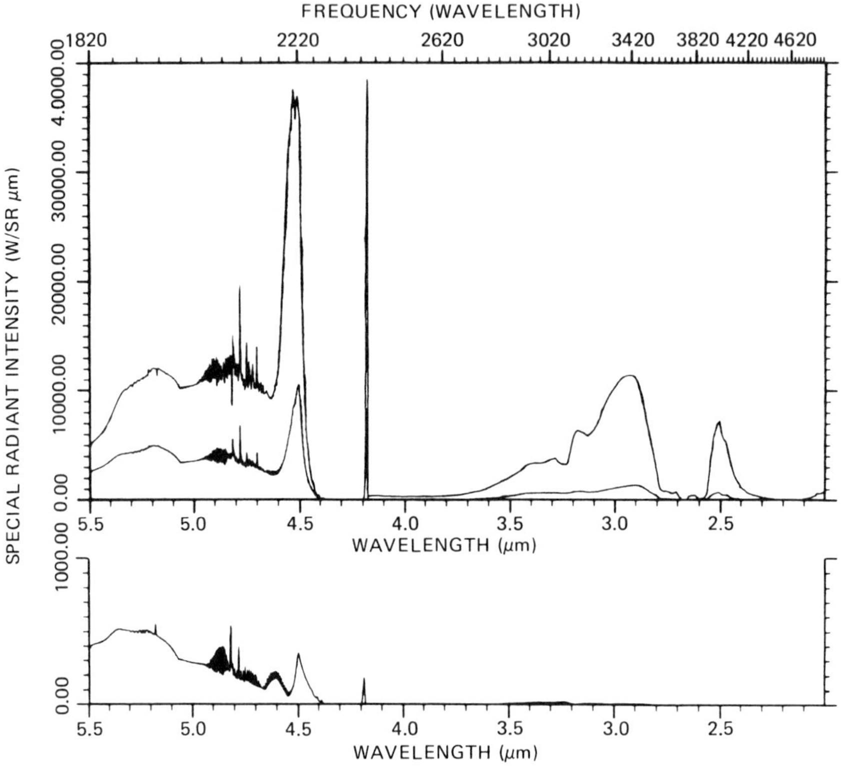

Fig. 21 Model stack plume spectra for core temperatures of 350, 550, and 700 K.

temperatures and spreading of population over more rotational states, the contrast between CO lines and the background decreases until at the highest temperature individual lines have almost disappeared. Also in Fig. 22, it can be seen that the increase in the CO_2 band which corresponds to termination on the ground state is much less than that for any of the neighboring hot bands.

Figures 23 and 24 show the effect of addition to the model plume of 5×10^{-4} mole fraction soot particles (the same concentration which was used as nominal in the model of the sooting flame). Unlike the effect on the relatively small flame, where molecular emission features were still clearly visible, this model smoke plume is large enough that even a relatively light particle loading

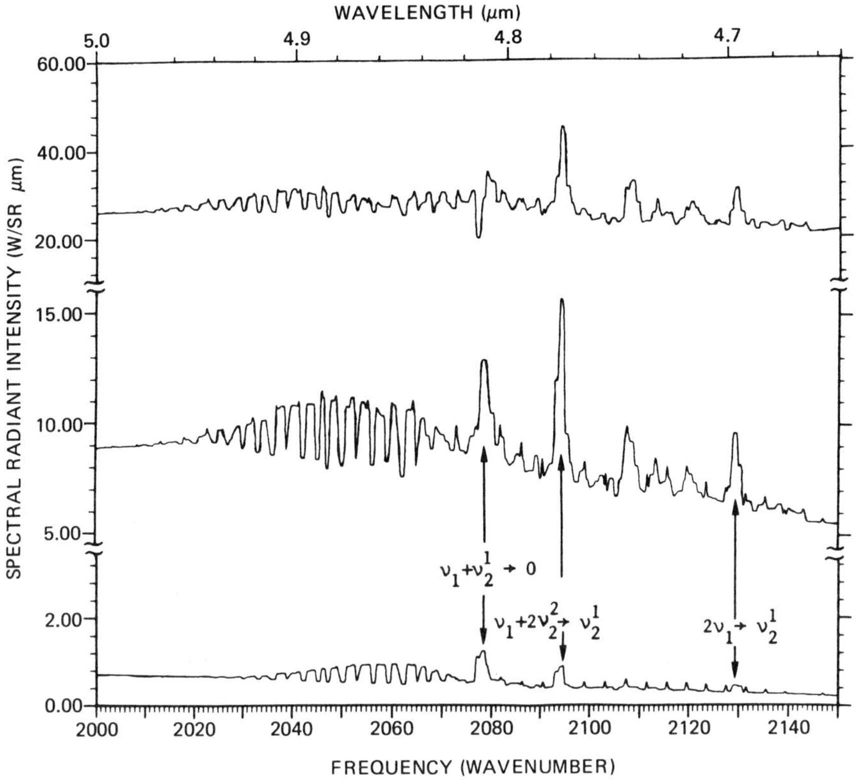

Fig. 22 Detail of hot CO and CO_2 emission region for model stack plume spectra for 350, 550, and 700 K core temperatures.

produces an opaque plume which resembles a backbody emitter with molecular absorption features superimposed. In particular, the CO_2 4.3-μm blue spike emission feature has disappeared into the particle continuum. In Fig. 23 it can be seen that both CO and CO_2 lines have reversed to absorption features.

Next we turn to size scaling, investigating the differences between the nominal prediction, corresponding to a stack radius of 3.0 m, and predictions for 1.0- and 5.0-m radii. The spectral shape for the smaller variation appears almost identical to that for the nominal case and is not shown in the comparison of Fig. 25, but numerical comparisons show that plume optical depth and temperature gradients actually affect the scaling. For example, the

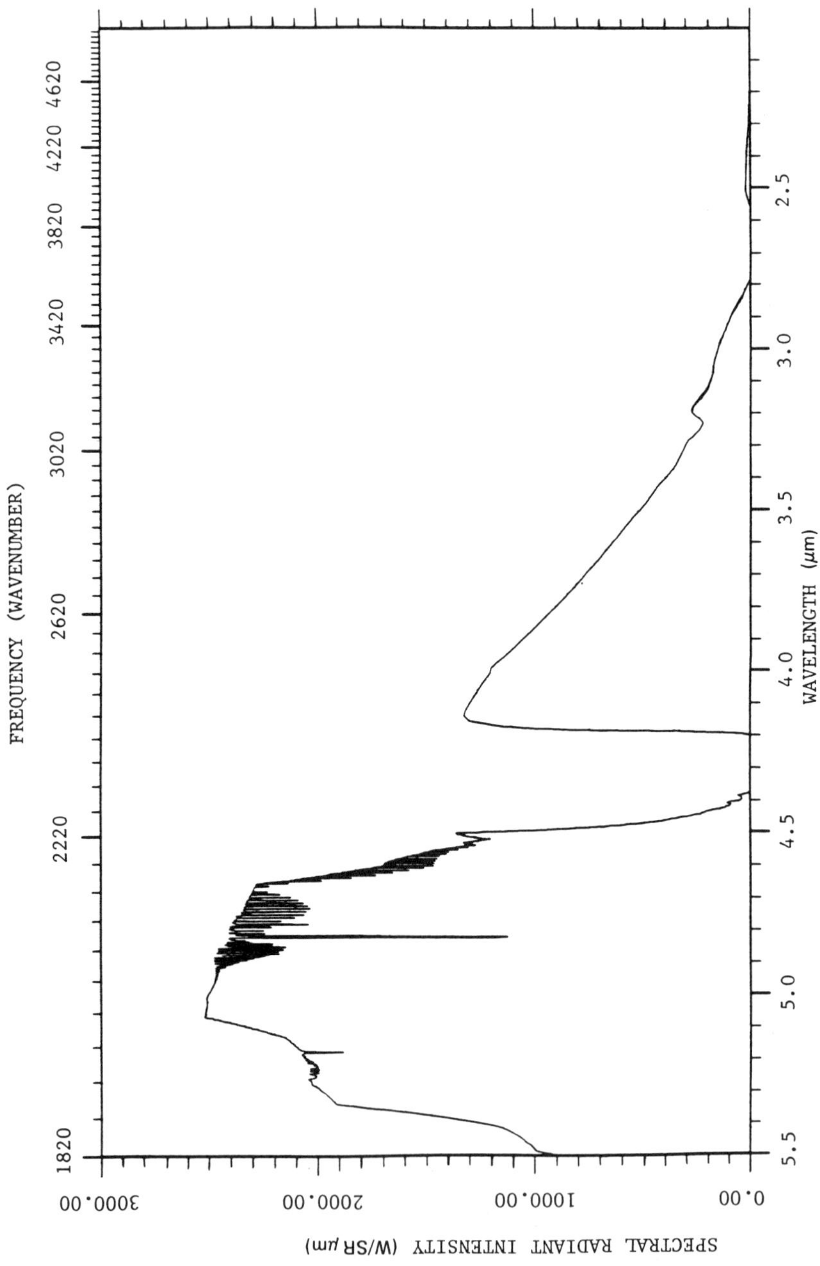

Fig. 23 Model stack plume spectrum including soot mole fraction of 5×10^{-4}.

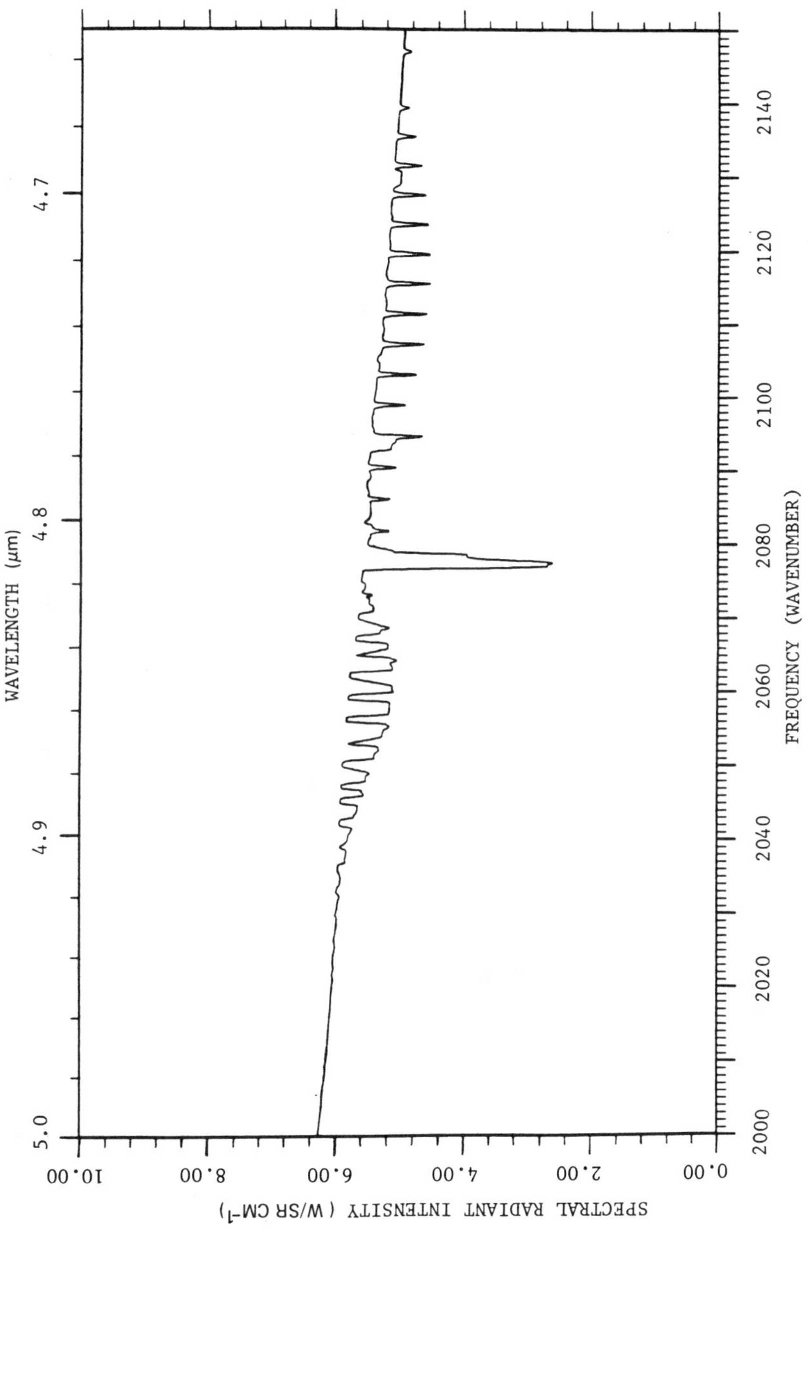

Fig. 24 Hot CO and CO_2 region detail of model stack plume spectrum including 5×10^{-4} soot mole fraction.

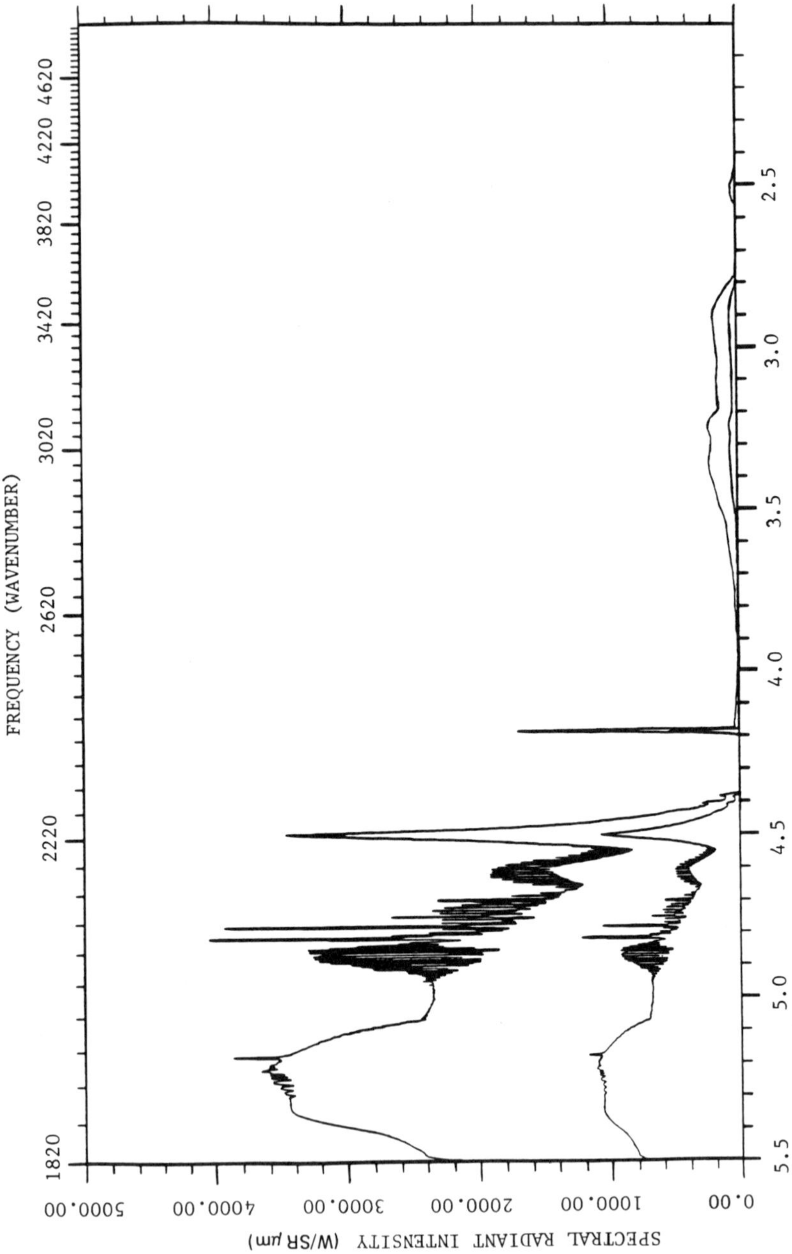

Fig. 25 Model plume spectra for plume radii of 300 (nominal) and 500 cm.

ratio of the short-wavelength water band peak turns out to be not $(0.33)^3$ but only $(0.33)^2 \times 0.67$ so that instead of scaling by volume, the increase in plume physical depth only provides half that increase in emission. The size of this effect is of course dependent on the spectral region, but even in the comparison between the nominal and larger plumes shown in Fig. 25, it is not sufficient to significantly distort the spectral shape. The chief reason for this scaling behavior is the milder temperature gradients occurring, in comparison to the flames studied above.

6 CONCLUSIONS

The first objective of this chapter was to give examples of the ability of simple infrared radiation models to accurately predict infrared emission characteristics of combustion sources at even higher spectral resolution than that currently measureable with Fourier transform interferometers. We have made quantitative comparisons which bear out the qualitative agreement shown here. The second objective was to illustrate the types of information about sources available in such FT-IR spectra. Only coupling with fairly detailed modeling will allow extraction of useful data for spectra. When a model is available, it can be used to convert absolute intensities or relative changes into information on several properties of interest.

Calculations with varying amounts of CO were able to indicate detection limits, as well as to exemplify the nonlinearity of emission with concentration induced by finite optical depths. Temperature variations illustrated spectral features which were seen to be especially temperature-sensitive, such as the 4.3-μm CO_2 red and blue spikes and the CO band wings. Addition of graybody radiators such as soot particles was seen to mask the hot gas regions, resulting (for large sources) in molecular spectral features reversing from emission to absorption. A similar process can take place with increases in source size with constant temperature profiles. Finally, differences in atmospheric transmission path length can have dramatic effects on the observed spectra, which are, however, separable from the variations in the source listed above.

In summary, the examples presented show that high resolution infrared observations coupled with models of the emission sources can provide useful information in a number of remote and automated sensing applications such as pollution monitoring and combustion efficiency control in both open and confined combustion systems.

ACKNOWLEDGMENT

We thank R. B. Lyons for contributions to the calculations and C. E. Kolb for useful discussions.

REFERENCES

Barnes, H. M., Jr., Herget, W. F., and Rollins, R. (1974). "Remote Sensing of SO_2 in Power Plant Plumes Using Ultraviolet Absorption and Infrared Emission Spectroscopy", *Analytical Methods Applied to Air Pollution Measurements* (R. K. Stevens and W. F. Herget, eds.), Ann Arbor Science Publishers, Ann Arbor, Michigan.

Barrett, R. E., Miller, S. E., and Locklin, D. W. (1973). "Field Investigation of Emissions from Combustion Equipment for Space Heating," Environmental Protection Agency Report EPA-R2-73-084a.

Bernstein, L. S. (1980). Band model parameters for the parallel bands of linear triatomic molecules, *J. Quant. Spectrosc. Radiat. Transfer, 23*: 157 and 169.

Bernstein, L. S., Wormhoudt, J., and Conant, J. A. (1979). "The Aerodyne Radiation Code (ARC): Physical Assumptions and Mathematical Approximations," Aerodyne Research, Inc., Report No. ARI-RR-183.

Boynton, F. P., Ludwig, C. B., and Thomson, A. (1968). Spectral emissivity of carbon particle clouds in rocket exhausts, *AIAA J., 6*: 865.

Cato, G. A., Buening, H. J., DeVivo, C. C., Morton, B. G., and Robinson, J. M. (1974). "Field Testing: Application of Combustion Modifications to Control Pollutant Emissions from Industrial Boilers," Environmental Protection Agency Report EPA-650/2-74-078-a.

Chan, S. H., Lin, C. C., and Low, M. J. D. (1973). Analysis of principles of remote sensing and characterization of stack gases by infrared spectroscopy, *Environ. Sci. Technol.: 7*, 424-427.

Chippett, S., and Gray, W. A. (1978). The size and optical properties of soot particles, *Comb. Flame, 31*: 149.

Dalzell, W. H., and Sarofim, A. F. (1969). Optical constants of soot and their application to heat-flux calculations, *J. Heat Transfer, 91*: 100.

Dash, S. M., and Pergament, H. S. (1978). "A Computational Model for the Prediction of Jet Entrainment in the Vicinity of Nozzle Boattails (The BOAT Code)," NASA CR-3075, Washington, D.C.

Dash, S. M., Pergament, H. S., Thorpe, R. D., and Abuchowski, S. A. (1980). "Operational Instructions for a Preliminary Version of the JANNAF Standard Plume Flowfield Model (SPF/1)," Aeronautical Research Associates of Princeton, Inc., A.R.A.P. Report No. 415, Princeton, New Jersey.

Dash, S. M., and Pergament, H. S. (1980). "A Computational System for the Analysis of Mixing/Chemical/Shock Processes in Supersonic Internal and Exhaust Plume Flowfields," AIAA Paper 80-1255, Hartford, Connecticut.

Environmental Protection Agency, (1973). *Compilation of Air Pollutant Emission Factors (Second Edition).*

Fishburne, E. S., and Pergament, H. S. (1978). The dynamics and radiant intensity of large hydrogen flames, 17th International Symposium on Combustion, Leeds, England, p. 1063.

Freeman, G. W., Ludwig, C. B., Malkmus, W., and Reed, R. (1979). "Development and Validation of Standardized Infrared Radiation Model (SIRRM) Gas/Particle Radiative Transfer Model," AFRPL-TR-79-55.

Gersh, M. E., Draper, J. S., Wormhoudt, J. C., Elgin, J. B., Kolb, C. E., Robertson, D. C., Bernstein, L. S., and Reardon, J. E. (1980). *JANNAF Handbook: Rocket Exhaust Plume Technology*, Chap. 3, CPIA Publication 263, Chemical Propulsion Information Agency, Laurel, Maryland.

Gladney, E. S., Small, J. A., and Gordon, G. E. (1976). Composition and size distribution of in-stack particulate material at a coal-fired power plant, *Atmospheric Environment, 10*: 1071.

Heck, P. H., and Ferguson, D. R. (1971). "Analytical Solution for Free Turbulent Mixing in Compressible Flows," AIAA Paper No. 71-74.

Herget, W. F. (1979). Overview of the environmental protection agency (EPA) programs for ground-based remote sensing of air pollution, *Proc. SPIE 195*: 87.

Herget, W. F. (1982). Remote and cross-stack measurement of stack gas concentrations using a mobile FT-IR system, *Appl. Opt. 21*: 635-641.

Herget, W. F., and Brasher, J. D. (1979). Remote measurement of gaseous pollutant concentrations using a mobile Fourier transform interferometer system, *Appl. Opt.*, *18*: 3404-3420.

Herget, W. F., and Brasher, J. D. (1980). Remote Fourier transform infrared air pollution studies, *Opt. Eng.*, *19*: 508-514.

Hommert, P. J., and Viskanta, R. (1978). High temperature gas diagnostics by spectral remote sensing, *Int. J. Heat Mass Transfer, 21*: 769.

Hottel, H. C., and Sarofim, A. F. (1967). *Radiative Transfer*, McGraw-Hill, New York.

Howarth, C. R., Foster, P. J., and Thring, M. W. (1966). "The Effect of Temperature on the Extinction of Radiation by Soot Particles," *Proceedings of the Third International Heat Transfer Conference*, American Institute of Chemical Engineers, New York, Vol. 5, p. 122.

Huffaker, R. M. (1966). "Inhomogeneous Radiant Heat Transfer from Saturn-Type Rocket Exhaust Plumes," AIAA Paper No. 66-652, Part I.

Huffaker, R. M. (1968). Current research on infrared radiation from rocket exhaust, *J. Quant. Spectrosc. Radiat. Transfer, 8*: 87.

Jensen, D. E., and Jones, G. A. (1981). Theoretical aspects of secondary combustion in rocket exhausts, *Combust. Flame, 41*: 71.

Jensen, D. E., Spaulding, D. B., Tatchell, D. G., and Wilson, A. S. (1979). Computation of structures of flames with recirculating flow and radial pressure gradients, *Combust. Flame, 34*: 309.

Jones, A. R. (1979). Scattering of electromagnetic radiation in particulate laden fluids, *Prog. Energy Combust. Sci.*, *5*: 73.

Low, M. J. D., and Clancy, F. K. (1967). Remote sensing and characterization of stack gases by infrared spectroscopy. An approach using multiple-scan interferometry, *Environ. Sci. Technol.*, *1*: 73-74.

Ludwig, C. B., Malkmus, W., Freeman, G. N., Slack, M., and Reed, R. (1981). "The Standardized Infrared Radiation Model (SIRMM)," AIAA Paper 81-1051, *AIAA 16th Thermophysics Conference*, Palo Alto, California.

Ludwig, C. B., Malkmus, W., Reardon, J. E., and Thompson, J. A. L., (1973). *Handbook of Infrared Radiation from Combustion Gases*, NASA SP 3080.

Menzies, R. T., and Shumate, M. S. (1974). Air pollution: remote detection of several pollutant gases with a laser heterodyne radiometer, *Science, 184*: 570-572.

Mikatarian, R. R., Kau, C. J., and Pergament, H. S. (1972). "A Fast Computer Program for Nonequilibrium Rocket Plume Predictions," AeroChem Research Laboratories, Report No. AFRPL-TR-72-94.

National Air Pollution Control Administration, Public Health Service, U.S. Department of Health, Education, and Welfare (1970). "Control Techniques for Carbon Monoxide Emissions from Stationary Sources."

Palmer, T. Y. (1976). Absorption by smoke particles of thermal radiation in large flames, *J. Fire and Flammability, 7*: 460.

Pearce, B. E. (1980). "The Naval Ship Exhaust Plume Signature Code," A.R.A.P. Report No. 428, Aeronautical Research Associates of Princeton.

Pergament, H. S., and Fishburne, E. S. (1978). Influence of buoyancy on turbulent hydrogen/air diffusion flames, *Combustion Science and Technology, 18*: 127–137.

Pergament, H. S., Fishburne, E. S., Abuchowski, S., Pearce, B. E., and Dash, S. M. (1979). "The Naval Weapons Center Target Signature Code," A.R.A.P. Report No. 380.

Pilat, M. J., and Ensor, D. S. (1971). Plume opacity and particulate mass concentration, *Atmospheric Environment, 5*: 209.

Prengle, H. W., Jr., Morgan, C. A., Fang, C. S., Huang, L.-K., Campani, P., and Wu, W. W. (1973). Infrared remote sensing and determination of pollutants in gas plumes, *Environ. Sci. Technol., 7*: 417-423.

Reardon, J. E. (1970). "Prediction of Radiation from Rocket Exhaust Plumes," AIAA Paper No. 70-841.

Rochelle, W. C. (1967). "Review of Thermal Radiation from Liquid and Solid Propellant Rocket Exhausts," NASA TM X-53579.

Rothman, L. S., Gamache, R. R., Barbe, A., Goldman, A., Gillis, J. R., Brown, L. R., Toth, R. A., Flaud, J. -M., and Camy-Peyret, C. (1983). AFGL atmospheric absorption line parameters compilation: 1982 edition, *Appl. Opt.*, *22*: 2247.

Scanlon, J. H., and Slack, M. W. (1975). "Band Model Computer Program for the Calculation of Infrared Radiation from Rocket Exhaust Plumes Containing HCl and HF," Grumman Research Corporation Memorandum RM-607.

Selby, J. E. A., and Reed, R. A. (1978). Atmospheric background and smoke stack measurements and modeling in the 3.8-5.5 μm region, *Proc. SPIE, 142*: 9.

Schulz, E. J., Engdahl, R. B., and Frankenberg, R. R. (1975). Submicron particles from a pulverized coal fired boiler, *Atmospheric Environment*, *9*: 111.

Shepherd, O., Hurd, A. G., Wattson, R. B., Smith, H. J. P., and Vanasse, G. A. (1981). Spectral measurements of stack effluents using a double-beam interferometer with background suppression, *Appl. Opt.*, *20*: 3972.

Siddall, R. G., and McGrath, I. A. (1963). The emissivity of luminous flames, *Ninth Symposium on Combustion*, Academic Press, New York, p. 102.

Stein, A. (1981). Heterodyne pyrometry in furnances, *Appl. Opt.*, *20*: 2871-2872.

Stone, C. W., and Tate, S. (1975). Report ASD/XR-TR-75, AF Aeronautical Systems Division, Wright-Patterson AFB, OH, January 1975, Avail. DTIC, AD B009668-70.

Stull, V. R., and Plass, G. N. (1960). Emissivity of dispersed carbon particles, *J. Opt. Soc. Am.*, *50*: 121.

Tanabe, L. H. (1975). "Determination of SO_2 Concentrations by Remote Sensing in the Infrared," General Dynamics Pomona Division Technical Memorandum 6-12SPH-430.

Tien, C. L., Doornink, D. G., and Rafferty, D. A. (1979). Attenuation of visible radiation by carbon smokes, *Comb. Sci. Tech.*, *6*: 55.

Wall, T. F., Lowe, A., Wibberley, L. J., Mai-Viet, R., and Gupta, R. P. (1981). Fly ash characteristics and radiative heat transfer in pulverized-coal-fired furnances, *Comb. Sci. Tech.*, *26*: 107.

Watson, G. H., and Lee, A. L. (1977). Thermal radiation model for solid rocket booster plumes, *J. Spacecraft, 14*: 641.

Young, S. J. (1975). Band model formulation for inhomogeneous optical paths, *J. Quant. Spectrosc. Radiat. Transfer, 15*: 483, and Young, S. J. (1975). "Band Model Parameters for the 2.7 μm Bands of H_2O and CO_2 in the 100 to 3000 K Temperature Range," SAMSO-TR-209, The Aerospace Corporation.

Young, S. J. (1981). "Retrieval of Flowfield Gas Temperature and Concentration in Low-Visibility Propellant Rocket Exhaust Plumes," AIAA Paper No. 81-1054.

2

Thermal and Momentum Structure of an Emerging Plume by Remote Sensing

H. WILLIAM PRENGLE, JR., UDAY MAHAGAOKAR,* and SHUN-KWOK TSE† *Department of Chemical Engineering, University of Houston–University Park, Houston, Texas*

Current affiliations:
*Shell Development Company, Houston, Texas
†Dupont Company, LaPorte, Texas

1 INTRODUCTION

In recent years considerable interest has developed in the monitoring of gaseous pollutants from stationary emission sources by the use of electrooptical techniques. Prengle et al. (1973) conducted preliminary investigations of remote sensing infrared radiometry–spectroscopy (IRRS) techniques using Fourier transform spectroscopic equipment to determine the temperature and concentration of pollutants in an emerging plume.

A subject of importance to meteorologists, engineers, and physicists studying emission spectral data and the behavior of gas pollutants is the thermal and momentum structure and turbulent fluctuations in a plume emerging into the atmosphere. Concerning the former, much of this book treats the information in hot gas infrared emission spectra, but absolute intensities contain information also. In using both absolute intensities and relative quantities, e.g., line peak ratios or band-to-continuum ratios, the ultimate accuracy is often limited by source fluctuations, which therefore require quantification.

In addition, the problem of determining the mechanism by which the energy in the hot gases dissipates and contributes to the dispersion of pollutants into the atmosphere is of interest. Several models have been proposed for the theoretical treatment of a continuous plume of buoyant gas moving through various types of atmospheres. A thorough review and critique of these theoretical models has been given by Briggs (1969), examining their usefulness for calculating the height to which plumes will rise under different conditions.

So far it has been extremely difficult to make thermal measurements on actual plumes of substantial size, and essentially impossible to follow the temperature gradients, axially and radially. Some measurements using a helicopter have been attempted, but these results have definite limitations, since the process of measurement alters the flow pattern of the plume (Davis et al., 1974; Tennessee Valley Authority, 1954; Bringfelt, 1968). Hence no body of information exists on the phenomena involved, except on emerging laboratory jets.

This chapter deals with the application of the aforementioned IRRS technique for temperature measurements on an operating power plant plume. Results consisting of temperatures, axial gradients, radial gradients, and turbulent temperature fluctuations within the plume are presented. The plume structure is deduced, and by *ab initio* calculations using a finite difference technique the experimental temperature data are interpreted. Also by a similarity hypothesis, the axial and radial decay of velocity is deduced.

2 EQUIPMENT SYSTEM

The equipment system used, consisting of a signal receiving and data collecting subsystem and a data processing subsystem, has been described in detail previously by Prengle et al. (1973). The Fourier transform spectrometer is a Michelson interferometer equipped with one fixed and one movable mirror covering the infrared region from 2.8 to 25 μm in one scan.

Temperature measurements of a remote target are based on the total energy received, which is a direct function of the peak-to-peak value of the maximum peak in the interferogram. In the early experiments, this value was read from a meter on the instrument control unit. The addition of a Biomation Model 610 transient recorder (Mahagaokar, 1976), capable of recording single events, or low repetition rate signals, for continuous CRT display and plotter records, greatly improved the accuracy. Triggering was done by the main signal with the level adjusted to display the part of the interferogram containing the peak.

2.1 Energy Transport Equation

Radiant energy emitted by a remote source is received and processed, and can be visualized as two processes: (a) the target side process—transport of the energy from the source to the receiver, and (b) the instrument side process—conversion of the energy received into intelligible signals.

This chapter will simply introduce a form of the equation for the transport of radiant energy. Chapters 3–5 will give more of the formalism of radiation transport, culminating in Chap. 5 with a full derivation. Here only a simple equation for the radiant power falling on the receiver in terms of properties of the source is required. This result will be presented after a review of nomenclature associated with radiative quantities.

Total *radiant power* emitted or absorbed is denoted by the symbol P and commonly has the units of watts (W). Spectral quantities will be indicated by a subscript $\bar{\nu}$, where $\bar{\nu}$ represents the frequency in wavenumber units (inverse wavelength, $1/\lambda$), or cm^{-1}. So, for instance, $P_{\bar{\nu}}$ is the spectral radiant power at wavenumber $\bar{\nu}$. The *radiant intensity* is the total radiant power per unit solid angle in a particular direction, having the units of W sr^{-1}. For a plane radiating surface, the projected area varies as the cosine of the angle with the normal to the surface. If the surface is also an isotropic radiator (meaning the radiance is independent of angle), then we obtain the radiant intensity by dividing by π, the integral of the cosine function over 2π solid angle. Multiplying the radiant intensity by the solid angle subtended by the receiver at the source gives the radiant power falling on the detector. This solid angle is calculated by ratioing the receiver area to the square of the distance between the source and the receiver. Dividing by the area of the receiver gives the *irradiance* H, in units of W m^{-2}. Dividing by the projected source area gives the *radiant emittance*, also in W m^{-2}.

This book is most concerned with the *radiance* at the source. The radiance is the power radiated per unit solid angle per unit projected area and is denoted here by N. Typical units are W m^{-2} sr^{-1}. Again, for an isotropic source like a blackbody surface or infinitesimally thin plane slab of gas, one first calculates the radiant emittance in W m^{-2}, then divides by π to obtain the radiance. Multiplying the radiance by the source projected area we again obtain the radiant intensity, in W sr^{-1}.

On the other hand, multiplying the radiance by the solid angle subtended by the source at the detector again gives the irradiance. This solid angle is computed by A_g/L^2, where A_g is the projected area of the volume of gas which is the source, and L is the source–receiver distance. Finally, multiplying the irradiance by the detector area (or that of the collecting lens or mirror, here represented by A_m) returns us to the radiant energy flux or radiant power falling on the receiver. This sequence from radiant emittance to radiance to irradiance to radiant power is contained in Eq. (1) for the transport of energy on the target side. With the notation that blackbody quantities have a zero superscript, so that for instance the radiance of a blackbody at temperature T is $N^0(T)$, we have

$$P_{\bar{\nu}} = (1 - e^{-\kappa_{\bar{\nu}} L_e})\left(\frac{A_m A_g}{L^2}\right) N^0_{\bar{\nu}}(T) e^{-\kappa^a_{\bar{\nu}} L_a} \tag{1}$$

A derivation of a similar relation from the equation of radiation transport is given in Chap. 5, culminating in Eq. (11).

The target side process involves transmission of source radiation to the receiver over a path L_a through an atmosphere whose spectral absorption coefficient is represented by $\kappa^{a}_{\bar{\nu}}$. It also involves the geometry of a three-dimensional plume which entails lengthy computations of the volumetric integral in the energy transfer equation. Fang and Prengle (1974) proposed a simplified method in which various plume geometries can be represented by an optical shape factor, to eliminate the volumetric integration.

Use of the equivalent path length L_e enables conversion of a three-dimensional remote emitting gas volume to a two-dimensional surface A_g, while maintaining the concentration-related variable $\kappa_{\bar{\nu}}$ in the equation. We can nondimensionalize L_e by $L_e \equiv \sigma L_c$, where σ is called the optical shape factor, and L_c is a reference length, appropriately defined for a given geometry. Here the appropriate L_c has been adopted as twice the plume radius, $2r_p$. A real gas volume from a stack can be approximated by two equal intersecting diameter cylindrical surfaces; the shape factor is given by

$$\sigma = -\left(\frac{1}{2\kappa_{\bar{\nu}} r_p}\right) \ln\left[1 - \frac{4}{\pi}\int_0^1 (1 - R^2)^{1/2}\left(1 - e^{-2\kappa_{\bar{\nu}} r_p (1-R^2)^{1/2}}\right) dR\right] \tag{2}$$

where the dimensionless radial coordinate R is r/r_p.

The instrument side process involves the Michelson interferometer where the incident radiation is split into two beams and sent to a fixed mirror and a moving mirror, respectively. The reflected beams meet at the beam splitter, and depending on the phase difference induced by the difference in path length or optical retardation, a resultant is sent to the detector. The detector frequency ν_d is related to the electromagnetic frequency ν by

$$\nu_d = (2\delta_0/ct)\nu \tag{3}$$

where δ_0 is half of the end-to-end interferometer optical retardation, and t is the time for the moving mirror to traverse this distance.

A mathematical treatment of interferometry has been presented by Bell (1972). The irradiance $H(\nu,\delta)$ for a particular frequency ν and optical retardation δ is given by

$$H(\nu,\delta) = c\varepsilon_0 E^2(\nu)[1 + \cos(2\pi\nu\delta)] \tag{4}$$

where ε_0 is the permittivity constant. Using a mean frequency $<\nu>$, the irradiance of the resultant wave is given by

$$H(\delta) = (1/<\nu>) \int_0^\infty H(\nu, \delta)\, d\nu \tag{5}$$

The interferogram, a plot of the function $[H(\delta) - \frac{1}{2}H(0)]$ versus δ, is given by

$$[H(\delta) - \frac{1}{2}H(0)] = (c\varepsilon_0/<\nu>) \int_0^\infty E^2(\nu) \cos(2\pi\nu\delta)\, d\nu \tag{6}$$

It follows that $c\varepsilon_0 E^2(\nu)$ is the irradiance $H(\nu)$ given by the Fourier cosine transform of the interferogram function

$$H(\nu) = <\nu> \int_0^\infty [H(\delta) - \frac{1}{2} H(0)] \cos(2\pi\nu\delta)\, d\delta \tag{7}$$

i.e., the Fourier cosine transform of the interferogram gives the spectrogram, which is a plot of irradiance versus frequency.

2.2 Temperature Determination

As was first demonstrated by Prengle et al. (1973), the temperature of a remote hot source can be determined by using the total flux method, wherein all component frequency intensities over the spectral range detected are added together. At zero retardation in the Michelson interferometer, both mirrors are equidistant from the beam splitter, $\delta = 0$, and Eq. (6) becomes

$$H(0) = (c\varepsilon_0/<\nu>)\, 2 \int_0^\infty E^2(\nu)\, d\nu \tag{8}$$

It is evident that at zero retardation all frequencies of the resultant wave are in phase and add to give a measure of the total flux. This condition is represented by the maximum amplitude in the interferogram which is proportional to the total energy in the incident radiation and hence directly related to the temperature of the target.

Substituting the emissivity $\varepsilon_{\bar{\nu}} = 1 - \exp(-k_{\bar{\nu}}^{a} L_e)$ and the atmospheric transmittance $\tau_a(\bar{\nu}) = \exp(-\kappa_{\bar{\nu}}^{a} L_a)$ into Eq. (1) and summing over all frequencies gives

$$H \equiv \left(\frac{P}{A_m}\right) = \left(\frac{A_g}{L_a^2}\right) \int_{\bar{\nu}_L}^{\bar{\nu}_H} \varepsilon_{\bar{\nu}} N_{\bar{\nu}}^{0}(T)\, \tau_a(\bar{\nu})\, d\bar{\nu} \tag{9}$$

The temperature signal is related to the irradiance by

$$\begin{array}{cccccc} S_T & = & R & G & A_m & H \\ (V) & = & (V/W) & (V/V) & (m^2) & (W/m^2) \end{array} \tag{10}$$

where R is the overall responsivity of the system and G is the instrument gain. Combining Eqs. (9) and (10) gives the temperature signal

$$S_T = (RGA_m A_g / L_a^2) \int_{\bar{\nu}_L}^{\bar{\nu}_H} \varepsilon_{\bar{\nu}} N_{\bar{\nu}}^{0}(T)\, \tau_a(\bar{\nu})\, d\bar{\nu} \tag{11}$$

2.3 Background Radiation

In addition to the radiative energy from the source, the collector will receive radiative energy from the background of the source which depends on many factors: sun's radiation, ambient temperature, atmospheric constituents behind the plume, and angle of elevation. The background radiation can be divided into two parts: S_{BI}, the contribution from the region around the target, and S_{BII}, radiation from behind the target passing through the target. Therefore, the total signal is the sum of the target contribution and background effects,

$$S_m = S_T + S_{BI} + S_{BII} = S_T(1 + S_{BI}/S_T + S_{BII}/S_T) \tag{12a}$$

$$= S_T(B_0 + S_{BII}/S_T) = BS_T \tag{12b}$$

where

$$B = B_0 + S_{BII}/S_T \tag{13}$$

Thus only if background contributions are small will the total signal be simply proportional to the temperature signal. Combining Eqs. (11) and (12b) and replacing the integral by average quantities, the transfer equation can be written in terms of the total signal,

$$S_m = (BRGA_m A_g / L^2)\,\varepsilon \bar{N}^0(T)\,\bar{\tau}_a(\bar{\nu}_H - \bar{\nu}_L) \tag{14}$$

The blackbody radiance used in Eq. (1), evaluated at a suitable mean wavenumber $\langle\bar{\nu}\rangle$, is given by

$$\bar{N}^0(T) = \frac{C_1\langle\bar{\nu}\rangle^3}{\pi}\left[\exp\left(\frac{C_2\langle\bar{\nu}\rangle}{T}\right) - 1\right]^{-1}$$

$$\cong \frac{C_1\langle\bar{\nu}\rangle^3}{\pi}\exp\left(\frac{-C_2\langle\bar{\nu}\rangle}{T}\right) \tag{15}$$

The area of the target can be approximated as $A_g = \pi r_p^2$, and the average atmospheric transmittance can be defined in terms of an average absorption coefficient as $\bar{\tau}_a = \exp(-\bar{\kappa}_a L_a)$. We can define the exponential in Eq. (15) as a normalized temperature signal

$$N_T \equiv \frac{\bar{N}^0(T)\pi}{C_1\langle\bar{\nu}\rangle^3} = \left[\frac{\exp(\bar{\kappa}_a L_a)}{B}\right]\left(\frac{1}{RA_m G}\right)\left[\frac{S_m r_p^2}{L_a^2 \varepsilon C_1 \langle\bar{\nu}\rangle^3 (\bar{\nu}_H - \bar{\nu}_L)}\right] \tag{16}$$

which contains (a) atmospheric background, (b) instrument, and (c) target factors. We can then define a temperature function of the form $a_0 + b_0/T$, which has the property of increasing as the temperature increases. Combining Eqs. (14), (15), and (16), the temperature function (Prengle et al., 1973) becomes

$$\eta = a_0 + b_0 \log_{10} N_T \tag{17}$$

a very useful linear relationship in log N_T, where b_0 is seen to be $2.303/C_2\langle\bar{\nu}\rangle$. As an absolute temperature measurement method, use of Eq. (17) requires good knowledge of the gas emissivity and the various geometric factors. Since these are prone to uncertainties, the spectroscopic methods described in later chapters have a clear advantage in accuracy. On the other hand, the speed at which measurements can be obtained together with good precision when plume chemistry is relatively unchanging makes it an excellent method for measuring turbulent temperature fluctuations.

2.4 Target Contributions

When a blackbody or graybody is scanned, the temperature signal is the sum of a continuous spectrum. On the other hand, the energy emitted by a hot gas plume in the infrared region is a discrete function of frequency. The total flux received by the instrument is, after subtracting background radiation, a sum of the contributions of various emitting species present in the plume. This radiation is a function of the concentration of each infrared active component, its emissivity at a characteristic frequency, the atmospheric transmittance, and the detector responsivity at that frequency. The major infrared active species in a combustion plume are CO_2 and H_2O with normal concentrations of 8–20%, while pollutant gases are present with a total concentration of $\leqslant 1.0\%$. Atmospheric CO_2 and H_2O absorb infrared radiation in the same frequency regions as these plume components produce emission. The two effects do not cancel, since emission occurs at a high temperature and consequently spans a larger frequency interval than the absorption, which occurs at ambient temperature. The net result, as shown by theoretical calculations (Mahagaokar, 1976), is that pollutant gases, though present in relatively small amounts (<1000 ppm), contribute significantly to the total flux (~15%). To prevent erroneous temperature estimates due to small changes in pollutant levels, the total flux is adjusted as follows:

$$S_T = S(CO_2 + H_2) + S\ (\text{pollutants}) \tag{18}$$

Defining

$$X \equiv S\ (\text{pollutants})/S\ (CO_2 + H_2O) \tag{19}$$

and using S $(CO_2 + H_2O)$ as the temperature signal, the normalized temperature signal is adjusted to

$$N_T = [\exp(\bar{\kappa}_a L_a)/B](1/RA_m G)[S_m r_p^2/L_a^2(1 + X)\varepsilon C_1 <\bar{\nu}>^3\ \Delta\bar{\nu}] \tag{20}$$

where $\Delta\bar{\nu} = \bar{\nu}_H - \bar{\nu}_L$ is the system bandwidth.

3 SYSTEM PARAMETER VALUES

The system parameters R, $\langle \bar{\nu} \rangle$, and B_0 were determined experimentally from measurements made with a 33.0-cm diameter graybody viewed at distances from 15.0 m to 120.0 m at three adjustable temperatures, 402 K, 454 K, and 550 K.

When the instrument is focused on a graybody, $S_{BII} = 0$ and $B = B_0$. Here B_0 depends on the size of the target relative to the field of view of the telescope and the temperature of the target relative to the ambient. These factors can be combined into two dimensionless quantities—the field of view number and the temperature ratio. Then

$$B_0 = f[(Lr_m/L_t r_p), (T/T_a)] \tag{21}$$

where r_m and L_t are the dimensions of the telescope and T_a is the ambient temperature; B_0 should satisfy the limiting relationship

$$\text{Lim } B_0 = 1 \quad \text{as} \quad (Lr_m/L_t r_p) \to 0 \tag{22}$$

A polynomial was obtained for $\log_{10} B_0$ using a multiple regression analysis of the experimental data, giving

$$\log_{10} B_0 = a_1(Lr_m/L_t r_p) + a_2(Lr_m/L_t r_p)^2 + b_1(Lr_m/L_t r_p)(T_a/T) \tag{23}$$

$a_1 = 0.5177 \times 10^{-2}$ standard deviation = 0.0862

$a_2 = 0.2368 \times 10^{-5}$

$b_1 = 0.1145 \times 10^{-1}$ average percent deviation = 3.1

Table 1 presents the system constants R and $\langle \bar{\nu} \rangle$ and the mean plume emissivity ε obtained from calibration runs on the power plant plume. $S_{BII}(\bar{\nu})$ was estimated by using a blackbody function at ambient temperature.

TABLE 1 System Parameters

Telescope constants:		
$A_m = 0.0475\ m^2$	$L_t = 0.546$ m	$r_m = 0.127$ m
Calibration constants:		
$<\bar{\nu}> = 1612.5\ cm^{-1}$	$\Delta\bar{\nu} = 3171.4\ cm^{-1}$	
ε(graybody) = 0.945	ε(plume) = 0.01816	
$\bar{\kappa}_a = 0.0023\ m^{-1}$	$b_0 = 0.9925 \times 10^{-3}$	
R = 56.4 V/W		

4 TEMPERATURE MEASUREMENTS

Data were obtained as successive interferograms recorded continuously for 5 to 7 min at each location on the plume. The peak value S_m of each interferogram, directly related to the momentary temperature of the plume, was determined by processing the recorded signal through a track and hold logic circuit. The temperatures obtained are characteristic of the plume area within the field of view, and successive values of S_m yield the fluctuating component of temperature at each location.

There are two basic configurations of the plume: in a calm atmosphere the plume rises vertically, but in the presence of a steady wind the plume is inclined. Measurements could not be taken in the latter case, since the plume would intermittently move out of the field of view. A summary of the processed data is presented in Tables 2, 3, 4, 5, and 6. Tables 2 and 3 show the axial temperature measurements for two runs, $<T_A>$, where the brackets denote time averaging and the subscript denotes area averaging, and T_{Arms}, the root mean squared fluctuating temperature components, shown as well, is the relative intensity of the temperature fluctuations $T_{Arms}/<T_A>$. Tables 4 and 5 are the radial measurements within one stack diameter, and the relative intensity of the fluctuations. Table 6 presents probability density values of the area-integrated temperature fluctuations $(T_A - <T_A>)/T_{Arms}$ at two positions along the plume path.

TABLE 2 Plume Temperature Data: Axial Profile in a Calm Atmosphere, Run 304[a]

Position	Gearhead angles θ_v (deg.) θ_h (deg.)	$z(m)$ z/D_j	$\langle T_A \rangle$ (K)	T_{Arms} (K)	$\frac{T_{Arms}}{\langle T_A \rangle}$
1	5.25 16.48	0.48 0.473	507	7.38	0.0146
2	5.5 16.48	0.774 0.763	502	8.05	0.0160
3	5.75 16.48	1.07 1.05	488	8.29	0.0170
4	6.0 16.48	1.36 1.34	472	10.2	0.0216
5	6.25 16.48	1.66 1.63	452	21.4	0.0474
6	6.75 16.48	2.25 2.21	436	18.4	0.0422
7	7.0 16.48	2.54 2.50	425	15.4	0.0362

[a]Furnace conditions: fuel flow rate = 0.44 m^3/sec, flue gas temperature = 467°F = 515 K, oxygen content in flue gas = 2.8%. Stack gas data: volumetric flow rate = 9.21 m^3/sec (at T = 507 K, P = 1 atm), mass flow rate = 6.11 kg/sec, velocity = 11.37 m/sec, Reynolds number = 2.737×10^5. Ambient conditions: T_a = 291.2 K, relative humidity = 29.0%.

5 PLUME STRUCTURE

Several previously published models (Schmidt, 1941; Rouse, Yih, and Humphreys, 1952; Sutton, 1953; Morton, Taylor, and Turner, 1956; Priestley and Ball, 1955) for the axial temperature decay of emerging plumes were used for comparison with the IRRS data. All models predict a large gradient near the stack exit; whereas the IRRS data indicate a striking concave down shape. After a distance of about one stack diameter, the gradient increases and

TABLE 3 Plume Temperature Data: Axial Profile in a Calm Atmosphere, Run 307[a]

Position	θ_v(deg.) θ_h(deg.)	$z(m)$ z/D_j	$\langle T_A \rangle$ (K)	$T_{A rms}$ (K)	$\frac{T_{A rms}}{\langle T_A \rangle}$
1	5.25 15.2	0.480 0.473	502	7.68	0.0153
2	5.60 15.2	0.890 0.877	500	8.66	0.0173
3	5.9 15.2	1.24 1.22	476	9.58	0.0201
4	6.2 15.2	1.59 1.57	455	17.1	0.0376
5	6.5 15.2	1.95 1.92	442	20.7	0.0468
6	6.8 15.2	2.30 2.26	430	17.8	0.0414
7	7.1 15.2	2.65 2.61	421	15.7	0.0373

[a]Furnace conditions: fuel flow rate = 0.424 m^3/sec, flue gas temperature = 464°F = 513 K, oxygen content in flue gas = 3.0%. Stack gas data: volumetric flow rate = 8.79 m^3/sec (at T = 502 K, P = 1 atm), mass flow rate = 5.90 kg/sec, velocity = 10.85 m/sec, Reynolds number = 2.639×10^5. Ambient conditions: T_a = 292.3 K, relative humidity = 32%.

then decays at a fairly uniform rate, behaving qualitatively as predicted by the models. The models assume that the radial distribution of temperature and velocity are approximately Gaussian in order to obtain an analytical solution of the conservation equations.

Tomich and Weger (1967) and others (Perry and Chilton, 1973; Kleinstein, 1964) showed that the thermal and momentum structure of a compressible, axially symmetric free jet can be described in terms of three flow regions, as shown in Fig. 1. The potential core region extends to the point where the turbulent mixing region reaches the jet axis. The transition region, where the spread of the jet in the radial direction and the axial decay of the

TABLE 4 Plume Temperature Data: Radial Profile in a Calm Atmosphere, Run 308[a]

Position	θ_v(deg.) θ_h(deg.)	$r(m)$ r/D	$\langle T_A \rangle$ (K)	T_{Arms} (K)	$\frac{T_{Arms}}{\langle T_A \rangle}$
1	5.35 15.0	0.364 0.359	406	13.8	0.0340
2	5.35 15.10	0.297 0.293	454	14.2	0.0313
3	5.35 15.25	0.209 0.206	475	11.2	0.0236
4	5.35 15.40	0.117 0.115	486	11.1	0.0228
5	5.35 15.50	0 0	491	9.36	0.0191
6	5.35 15.60	−0.117 −0.115	488	11.6	0.0238
7	5.35 15.78	−0.210 −0.207	477	13.4	0.0281
8	5.35 15.96	−0.286 −0.282	451	13.2	0.0293
9	5.35 16.05	−0.373 −0.368	401	15.6	0.0389

[a]Furnace conditions: fuel flow rate = 0.354 m^3/sec, flue gas temperature = 449°F = 505 K, oxygen content in flue gas = 3.1%. Stack gas data: volumetric flow rate = 7.17 m^3/sec (at T = 491 K, P = 1 atm), mass flow rate = 4.91 kg/sec, velocity = 8.84 m/sec, Reynolds number = 2.199×10^5. Ambient conditions: T_a = 296.2 K, relative humidity = 59%.

center-line properties are very rapid, is followed by the fully developed region, where the radial spread is nearly linear with respect to the axial distance. The radial distributions of velocity and excess temperature in the fully developed region are approximately Gaussian.

TABLE 5 Plume Temperature Data: Radial Profile in a Calm Atmosphere, Run 313[a]

Position	θ_v(deg.) θ_h(deg.)	$r(m)$ r/D_j	$\langle T_A \rangle$ (K)	T_{Arms} (K)	$\frac{T_{Arms}}{\langle T_A \rangle}$
1	5.35 15.0	0.364 0.359	400	15.2	0.0380
2	5.35 15.10	0.297 0.293	457	13.1	0.0287
3	5.35 15.25	0.209 0.206	479	12.9	0.0269
4	5.35 15.40	0.117 0.115	495	10.5	0.0212
5	5.35 15.50	0 0	500	9.14	0.0183
6	5.35 15.60	−0.117 −0.115	490	10.6	0.0216
7	5.35 15.75	−0.209 −0.206	477	12.3	0.0258
8	5.35 15.90	−0.297 −0.293	453	14.1	0.0311
9	5.35 16.0	−0.364 −0.359	403	14.2	0.0352

[a]Furnace conditons: fuel flow rate = 0.377 m^3/sec, flue gas temperature = 438°F = 499 K, oxygen content in flue gas = 2.8%. Stack gas data: volumetric flow rate = 7.78 m^3/sec (at T = 500 K, P = 1 atm), mass flow rate = 5.24 kg/sec, Reynolds number = 2.346×10^5. Ambient conditons: T_a = 297.2 K, relative humidity = 42%.

Based on the above description, it is assumed that the thermal and momentum structure of an emerging plume should resemble that of a free jet.

TABLE 6 Plume Temperature Data: Probability Density Function of Temperature Fluctuations, Run 304

At $z = 0.48$ m, $\langle T_A \rangle = 507$ K, $T_{Arms} = 7.38$ K.			At $z = 2.54$ m, $\langle T_A \rangle = 425$ K, $T_{Arms} = 15.4$.		
T_A (K)	$\frac{T_A - \langle T_A \rangle}{T_{Arms}}$	Probability density	T_A	$\frac{T_A - \langle T_A \rangle}{T_{Arms}}$	Probability density
489	− 2.40	0.050	397	− 1.80	0.015
491	− 2.10	0.075	402	− 1.50	0.28
497	− 1.20	0.22	406	− 1.20	0.47
501	− 0.80	0.30	413	− 0.75	0.62
503	− 0.50	0.37	418	− 0.40	0.47
509	0.30	0.43	423	− 0.10	0.30
511	0.50	0.33	430	0.30	0.23
514	0.90	0.22	440	1.00	0.08
518	1.50	0.075	446	1.40	0.07
520	1.80	0.02	453	1.80	0.06
			463	2.5	0.025

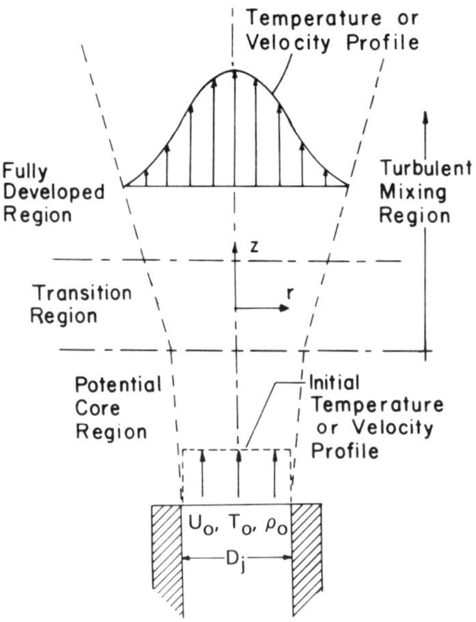

Fig. 1 Free jet and plume structure.

6 METHOD OF ANALYSIS

As a simulation of the IRRS temperature measurements, a computational technique based on finite-difference solution of the boundary layer forms for conservation of mass, momentum, and energy was developed, in order to calculate, *ab initio*, both the axial and radial distribution of temperature and velocity. The technique had been used previously by Tomich and Weger (1967) for a high velocity turbulent free jet and showed good agreement with experimental data. The finite-difference solution of the equations is a fully implicit method used by Abbott (1946). No transformations are made; solutions are obtained for free jet injection into a medium at rest, where the initial free jet velocity and temperature profiles are assumed to be step functions. Dynamic eddy transfer coefficients of momentum $\rho\varepsilon_v$ and heat flux k_T are defined for the turbulent compressible flow, which are modifications of those introduced by Kleinstein (1964) and Tomich and Weger (1967) and are functions of axial positions only.

Equations were derived from conservation of mass, momentum, and energy in the turbulent, compressible, axially symmetric, steady state free jet using the coordinate system shown in Fig. 1, as follows:

conservation of mass,

$$\frac{\partial}{\partial z}(\rho u_z) + \frac{1}{r}\frac{\partial}{\partial r}(r\rho u_r) = 0 \tag{24}$$

conservation of momentum,

$$\rho u_z \frac{\partial u_z}{\partial z} + \rho u_r \frac{\partial u_z}{\partial r} = \frac{1}{r}\frac{\partial}{\partial r}\left(r\rho\varepsilon_\nu \frac{\partial u_z}{\partial z}\right) + (\rho_a - \rho)g \tag{25}$$

and conservation of energy,

$$\rho u_z C_p \frac{\partial T}{\partial z} + \rho u_r C_p \frac{\partial T}{\partial r} = \frac{1}{r}\frac{\partial}{\partial r}\left(rk_T \frac{\partial T}{\partial r}\right) + \rho\varepsilon_\nu\left(\frac{\partial u_z}{\partial r}\right)^2 \tag{26}$$

In the above equations, r and z are radial and axial coordinates, u_r and u_z the velocities, ρ and ρ_0 the plume and ambient densities, ε_ν is the eddy kinematic viscosity, g the acceleration of gravity, C_p the heat capacity at constant pressure, and k_T the eddy thermal conductivity.

All densities, velocities, and temperature have been time-smoothed for turbulent flow; the heat capacity and the static pressure are assumed to be constant. With r_j representing the jet or stack radius, half of the D_j shown in Fig. 1, the boundary conditions for free jet flow are

at $z = 0$

$$u_z = u_0 \quad \text{for} \quad r < r_j$$

$$T = T_0$$

$$u_r = 0$$

$$u_z = 0 \quad \text{for} \quad r \geqslant r_j$$

$$T = T_a$$

$$u_r = 0$$

at $r = 0$

$$\frac{\partial u_z}{\partial r} = 0 \quad \text{for all} \quad z$$

$$\frac{\partial T}{\partial r} = 0$$

at $r \to \infty$

$$u_z = 0 \quad \text{for all} \quad z$$

$$T = T_a$$

$$u_r = 0$$

Equations (24)-(26) are made dimensionless by

$Z \equiv z/r_j$, $R \equiv r/r_j$, $\mathsf{P} \equiv \rho/\rho_0$, $U_z \equiv u_z/u_0$, $U_R \equiv u_r/u_0$,

$\mathsf{T} \equiv (T - T_a)/(T_0 - T_a)$

u_0, T_0, and ρ_0 are initial values at $z = 0$. Then

$$\frac{\partial}{\partial Z}(\mathsf{P}U_z) + \frac{1}{R}\frac{1}{\partial R}(R\mathsf{P}U_R) = 0 \tag{27}$$

$$\mathsf{P}U_z \frac{\partial U_z}{\partial Z} + \mathsf{P}U_R \frac{\partial U_z}{\partial R} = \frac{1}{u_0 r_j}\frac{1}{R}\frac{\partial}{\partial R}\left(R\mathsf{P}\varepsilon_\nu \frac{\partial U_z}{\partial R}\right) + \frac{r_j g}{u_0^2}(\mathsf{P}_a - \mathsf{P}) \tag{28}$$

$$\mathsf{P}U_z \frac{\partial \mathsf{T}}{\partial Z} + \mathsf{P}U_R \frac{\partial \mathsf{T}}{\partial R} = \frac{1}{\rho_0 u_0 r_j}\frac{1}{R}\frac{\partial}{\partial R}\left(R\frac{k_T}{C_p}\frac{\partial \mathsf{T}}{\partial R}\right) + \frac{u_0}{r_j(T_0 - T_a)C_p}\mathsf{P}\varepsilon_\nu\left(\frac{\partial U_z}{\partial R}\right)^2 \tag{29}$$

and the transformed boundary conditions are

at $Z = 0$

$$U_z = 1.0 \quad \text{for} \quad R < 1$$

$$T = 1.0$$

$$U_R = 0$$

$$U_z = 0 \quad \text{for} \quad R \geq 1$$

$$T = 0$$

$$U_R = 0$$

at $R = 0$

$$\frac{\partial U_z}{\partial R} = 0 \quad \text{for all} \quad Z$$

$$\frac{\partial T}{\partial R} = 0$$

$$U_R = 0$$

at $R \to \infty$

$$U_z = 0 \quad \text{for all} \quad Z$$

$$T = 0$$

$$U_R = 0$$

In this work, Tomich and Weger's eddy transfer coefficient, which was semiempirically deduced by Kleinstein, was modified slightly to provide a better correlation with experimental data. The eddy transfer coefficient for momentum is

$$\rho\varepsilon_\nu = F(M) C_p \rho_a^{1/2} \rho_m \rho_0 u_0 r_j f_\nu(Z) \tag{30}$$

where

$$F(M) = 0.0168 - 0.155\, M + 0.388\, M^2 - 0.267\, M^3$$

$$f_\nu(Z) = 0.2 \quad \text{for} \quad Z \leq Z_{c\nu}$$

$$f_v(Z) = 1.0 \quad \text{for} \quad Z > Z_{cv}$$

$$Z_{cv} = 17.5\left(\frac{T_0}{T_A}\right)^{-1.13} M^{0.430}$$

where P_m is $\rho_{r=0}/\rho_0$ and M is the Mach number. The eddy thermal conductivity is

$$k_T = F(M) C_p P_a^{1/2} P_m \rho_0 u_0 r_j f_t(Z) / N_{Pr} \tag{31}$$

where

$$f_t(Z) = 0.2 \quad \text{for} \quad Z \leqslant Z_{ct}$$

$$f_t(Z) = 1.0 \quad \text{for} \quad Z > Z_{ct}$$

$$Z_{ct} = 14.4 \frac{T_0}{T_a}^{-1.16} M^{0.455}$$

and the Prandtl number N_{Pr} is given by

$$N_{Pr} = 1.0 \quad \text{for} \quad Z \leqslant Z_{ct}$$

$$N_{Pr} = 0.175 \quad \text{for} \quad Z > Z_{ct}$$

These transfer coefficients differ from those of Tomich and Weger's analysis by adjusting the functions $F(M)$, Z_{cv}, and Z_{ct}. $F(M)$ has been extended to lower velocities using the remote sensing data, and by least squares $F(M)$ was obtained as a polynomial in Mach number. No buoyancy effect was considered in Tomich and Weger's conservation equations, and the functions of Z_{ct} and Z_{cv} were only dependent on the exit to ambient temperature ratio. The values of Z_{ct} and Z_{cv} were readjusted to fit the experimental data; a linear regression method was used to correlate Z_{ct} and Z_{cv} as a function of the temperature ratio and the Mach number.

Equations (30) and (31) were substituted into Eqs. (28) and (29),

$$\underline{\mathsf{P}U_z} \frac{\partial U_z}{\partial Z} + \underline{\mathsf{P}U_R} \frac{\partial U_R}{\partial R} = A_\nu \frac{1}{R} \frac{\partial}{\partial R}\left(R \frac{\partial U_z}{\partial R}\right) + \frac{r_j g}{u_0^2}(\mathsf{P}_a - \mathsf{P}) \tag{32}$$

where

$$A_\nu = \frac{\mathsf{P}\varepsilon_\nu}{u_0 r_j} = F(M)\mathsf{P}_a^{1/2}\mathsf{P}_m f_\nu(Z)$$

and

$$\underline{\mathsf{P}U_z} \frac{\partial \mathsf{T}}{\partial Z} + \underline{\mathsf{P}U_R} \frac{\partial \mathsf{T}}{\partial R} = A_t \frac{1}{R} \frac{\partial}{\partial R}\left(R \frac{\partial \mathsf{T}}{\partial R}\right) + \mathrm{N_{Ec}} A_\nu \left(\frac{\partial u_z}{\partial R^2}\right)^2 \tag{33}$$

where

$$A_t \equiv \frac{k_T}{C_p \rho_0 u_0 D} = F(M)\mathsf{P}_a^{1/2}\mathsf{P}_m f_t(Z)/\mathrm{N_{Pr}}$$

and the Eckert number is

$$\mathrm{N_{Ec}} \equiv \frac{u_0^2}{C_p(T_0 - T_a)}$$

In applying the finite-difference technique, the following approximations for derivatives at $(Z + \Delta Z, R)$ were used:

$$\frac{\partial U_z}{\partial Z} = \frac{U_z(Z + \Delta Z, R) - U_z(z, R)}{\Delta Z} \tag{34}$$

$$\frac{\partial U_z}{\partial R} = \frac{U_z(Z + \Delta Z, R + \Delta R) - U_z(Z + \Delta Z, R - \Delta R)}{2\Delta R} \tag{35}$$

$$\frac{\partial^2 U_z}{\partial R^2} = \frac{U_z(Z + \Delta Z, R + \Delta R) - 2U_z(Z + \Delta Z, R) + U_z(Z + \Delta Z, R - \Delta R)}{\Delta R^2} \tag{36}$$

and similar expressions were used for the derivatives of radial velocity, temperature, and density. The finite difference approximations were substituted into Eqs. (27), (32), and (33), which, along with the boundary conditions, comprise the system of nonlinear algebraic equations to be solved at each point of the finite-difference network. The system of algebraic equations was linearized by taking the coefficients (shown as the underlined coefficients in Eqs. (32) and (33)) of the nonlinear terms at their known value on the previous line of the network or at Z. The resulting system of simultaneous, linear algebraic equations was solved by using the Gaussian elimination method (Carnahan, Luther, and Wilkes, 1969) for initial values of U_Z, U_R, and T at each point of the line $Z + \Delta Z$. The computational accuracy was then improved by an iterative procedure, and the solution was carried downstream to $Z + 2\Delta Z$.

7 DATA ANALYSIS

The results of the numerical calculations are compared with the remote sensing data and the experimental free jet data of Corrsin and Uberoi (1950) and Tomich and Weger (1967). Axial direction comparisons are shown in Figs. 2, 3, and 4 and radial direction in Figs. 5, 6, and 7.

Simulation of the remote sensing temperature was made by taking the average of the point values from the finite-difference calculations in the area representing the field of view. Results from the numerical calculations agree well with the experimental results. Comparing Figs. 2, 3, and 4 shows that the axial temperature and velocity decay faster as the initial velocity decreases and as the temperature ratio T_0/T_a increases. The radial temperature and velocity distributions in Figs. 6 and 7 show that the radial profiles are Gaussian in shape in the fully developed region, and the characteristic radius describing the temperature distribution is greater than that for the velocity distribution. From the numerical calculations, the radial profiles of temperature and velocity at short distance change gradually from a flat profile to the Gaussian-like profile, as shown in Fig. 8. The edges of the profile are smoothing out in the Gaussian shape.

It will be noted that the axial temperature decays faster than the axial velocity. Since the thermal and momentum dissipation is mainly by convection, and for stack gas or hot air, the Prandtl number is less than unity, a more rapid decay in temperature is to be expected.

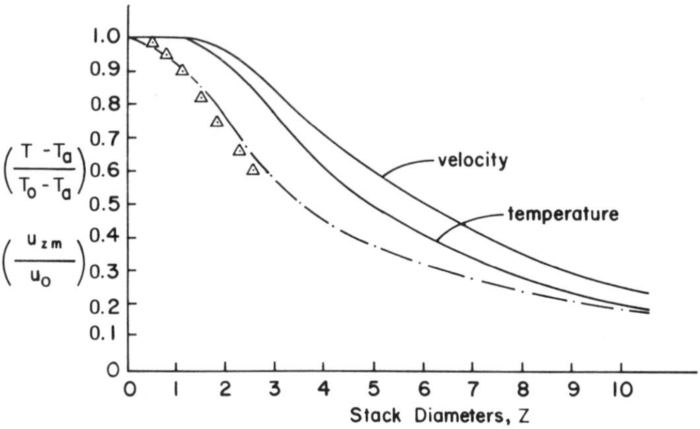

Fig. 2 Axial decay comparison: experimental data versus analysis. Δ, IRRS temperature data; u_0 = 37.3 ft/sec; T_0/T_a = 1.75; D_j = 3.33 ft; –, finite-difference analysis; •–•, average temperature of field of view from finite difference.

8 TEMPERATURE FLUCTUATIONS

The relative intensity of the area-integrated temperature fluctuations as a function of axial distance and radial positions is shown in Figs. 9 and 10. They are probably the most significant and revealing data obtained. From Fig. 9, the three flow regions of the emerging plume can be distinguished. The axial relative intensity varies from 1.5 to 5.0%, which is slightly lower than obtained by Corrsin and Uberoi (1950), for velocity and temperature fluctuations in a free jet—the latter were as high as 10%. The region close to the stack reflects the self-generated turbulence within the plume as shown by the lower relative intensities. After a distance of approximately 1.2 stack diameters, the effect of entrainment becomes appreciable, and additional turbulence is generated as shown by a rapid increase in relative intensity. The temperature core region length, determined by finite difference calculation, is about 1.2 diameters. After about 1.8 diameters, the turbulence decays slowly due to reduction in the rate of entrainment caused by a decay in the axial flow velocity.

For the radial profile of relative intensity at an axial distance of 0.6 stack diameters, as shown in Fig. 10, the temperature fluctuation intensity increases from the plume center line to the

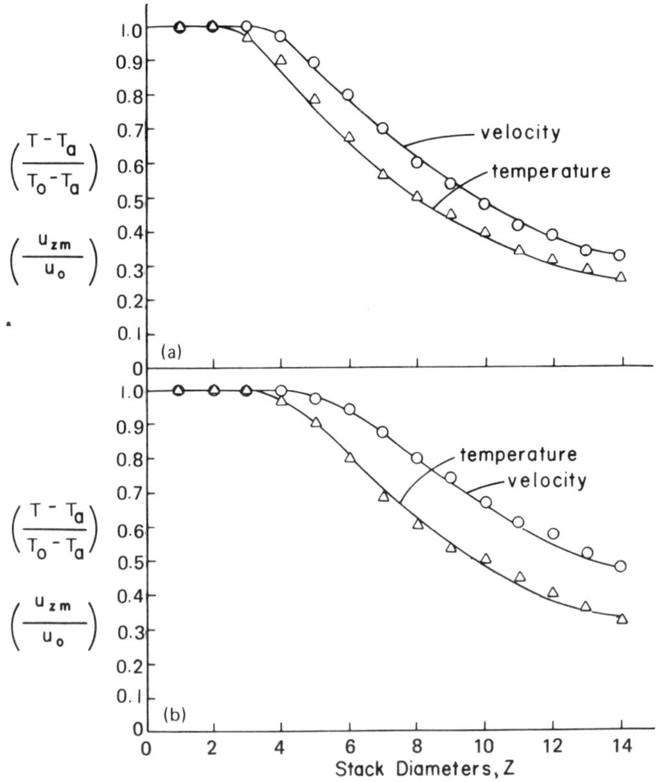

Fig. 3 Axial decay comparison: experimental data versus analysis. (a) –, Finite-difference analysis; $T_0/T_a = 2.02$. (b) –, Finite-difference analysis; $T_0/T_a = 1.05$.

outer edge. The behavior is similar to the radial profiles obtained by Becker, Rosenweig, and Gwozdz (1966) for the relative intensity of point concentration fluctuations in fully developed pipe flow and by Rosenweig, Hottel, and Williams (1961) for concentration fluctuations in a round turbulent jet.

Probability density functions of the area-integrated temperature fluctuations in two regions along the path of the plume are shown in Fig. 11. In the region close to the source, the function is Gaussian-like with a slight skewing to the left, which makes the most probable value occur at temperatures lower than the mean value. This is indicative of the fact that the plume has an

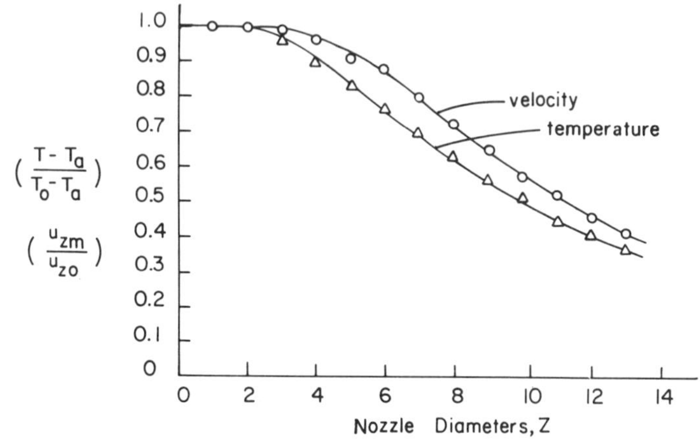

Fig. 4 Axial decay comparison: experimental data versus analysis (from Tomich and Weger, 1967). ○, △, Experimental data; u_0 = 687 ft/sec; T_0/T_a = 2.72; D_j = 0.375 in.

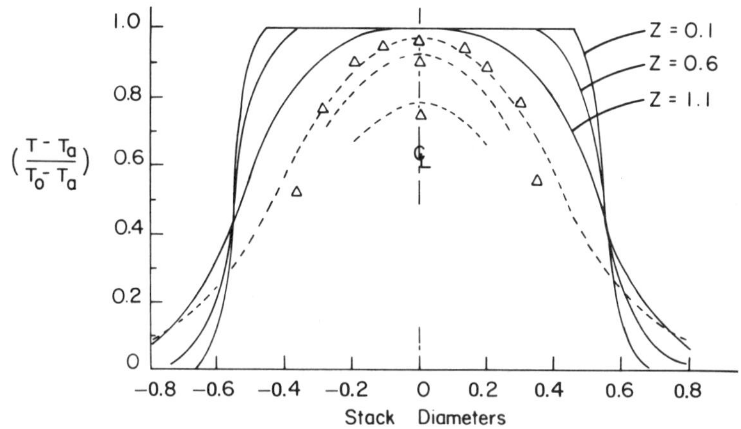

Fig. 5 Radial distribution comparison: experimental data versus analysis. Finite-difference analysis: –, point values; – – –, average value of field of view; △, IRRS temperature data; u_0 = 29 ft/sec; T_0/T_a = 1.68; D_j = 3.33 ft.

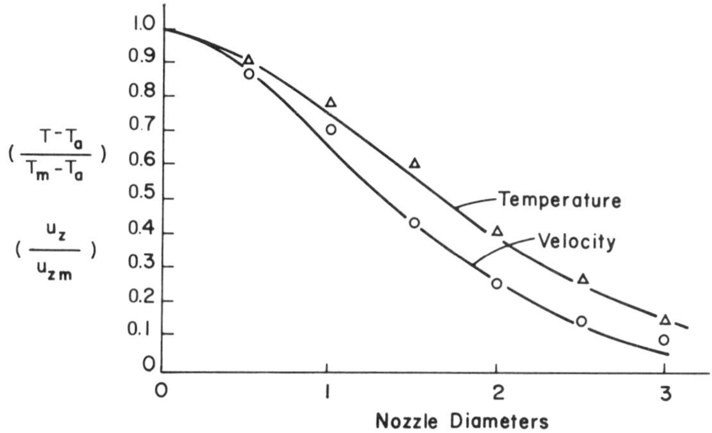

Fig. 6 Radial distribution comparison: experimental data versus analysis (from Corrsin and Uberoi, 1950). –, Finite-difference analysis, at $Z = 15$. ○, △, Experimental data; $u_0 = 78$ ft/sec; $T_0/T_a = 2.02$; $D_j = 1.0$ in.

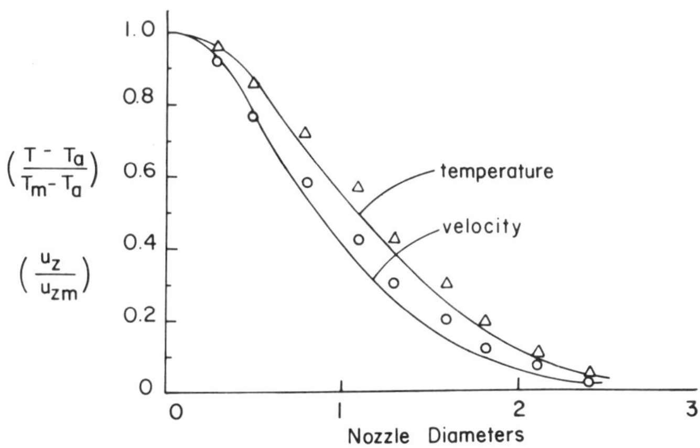

Fig. 7 Radial distribution comparison: experimental data versus analysis (from Tomich and Weger, 1967). –, Finite-difference analysis, at $Z = 11$. ○, △, Experimental data; $u_0 = 687$ ft/sec; $T_0/T_a = 2.72$; $D_j = 0.375$ in.

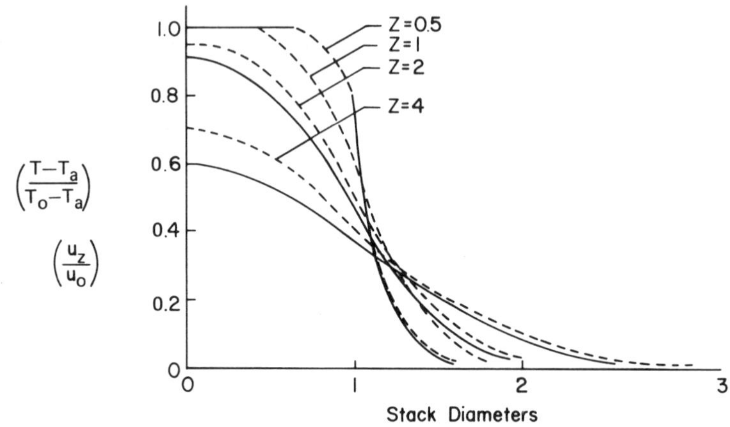

Fig. 8 Radial distributions at short distances from finite-difference analysis. –, Temperature; – – –, velocity; u_0 = 37.3 ft/sec; T_0/T_a = 1.75; D_j = 3.33 ft.

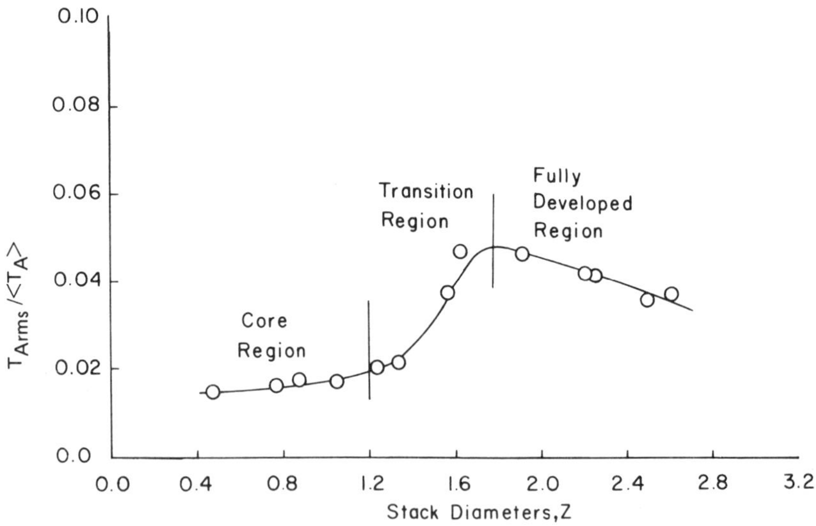

Fig. 9 Axial profile: relative intensity of area-integrated temperature fluctuations.

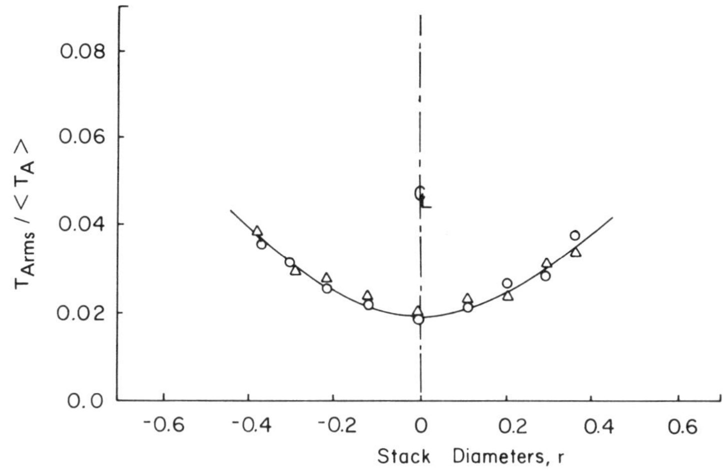

Fig. 10 Radial profile: relative intensity of area-integrated temperature fluctuations. ○, Run 313; △, run 308; $Z = 0.60$.

initial buoyancy as it just rises out of the stack so that a volume of higher temperature air is pushed up to a region just above the stack. At a distance of about 2.54 stack diameters, the function has a definite skewing to the left with the most probable value occurring at temperatures lower than the mean value. The effect of turbulent air mixing and entrainment is fairly strong in this region, which tends to lower the plume temperature, while there is also a finite but low probability of temperature being considerably higher than the mean value.

9 CONCLUSIONS AND DISCUSSION

1. Remote sensing IRRS techniques, with no disturbance of flow pattern, can be used to determine the temperature, axial and radial gradients, and fluctuations of a plume emerging into the atmosphere.

2. By appropriate analysis of the temperature data and a similarity hypothesis, the thermal and momentum structure of an emerging plume can be determined.

3. The thermal and momentum structure of an emerging plume resembles that of a laboratory scale free jet, described by core, transition, and fully developed regions.

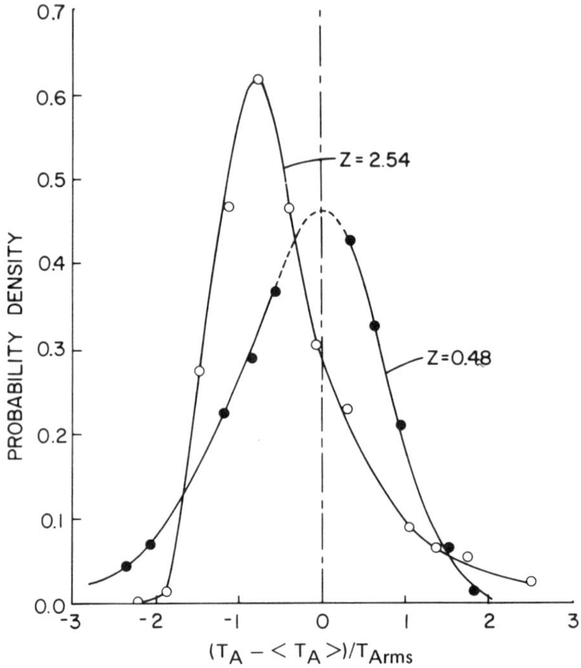

Fig. 11 Probability density of area-integrated temperature fluctuations. Regions centered at $Z = 2.54$ and $Z = 0.48$.

The remote sensing data at short axial distance indicate the dominant factor is the on-rush of high temperature and velocity gases into a relatively quiescent mass of low temperature gas, and the thermal and momentum structure resembles fully developed pipe flow in the core region. After the core, the temperature decays at a fairly uniform rate.

4. The existence of core, transition, and fully developed regions is confirmed experimentally by the relative intensity temperature fluctuation and probability density data.

These data indicate that the temperature decay is caused by entrainment and mixing between the plume and surrounding air, which is a result of mechanical turbulence generated by the velocity shear between the stack gas and the air.

5. For *ab initio* calculations, a finite difference technique with introduction of turbulent transfer coefficients into the conserva-

tion equations provides a good simulation of the plume and eliminates the need to make simplifying assumptions required in previous solutions.

Starting with pipe flow conditions as initial velocity and temperature distributions, calculations indicate that the edges of the initial radial profile become rounded and the flat profile gradually changes to Gaussian-like. Axial decay in the first-to-second stack diameter is almost negligible—the distance depending on the initial velocity and temperature ratio. As the radial profile approaches Gaussian, the axial decay becomes very fast and is greatest for the next one to two diameters; then the gradient decreases gradually as the axial distance increases.

6. The axial temperature and velocity decay faster as the initial temperature ratio increases and as the initial velocity decreases.

7. The radial temperature and velocity distributions are approximately Gaussian in the fully developed region, but the characteristic radius describing the temperature distribution is greater than that for the velocity.

The Gaussian-like profiles for velocity and temperature do not correspond, since the Prandtl number, the ratio of the momentum diffusivity to thermal diffusivity, is less than unity for gases.

REFERENCES

Abbott, M. S. (1964). Axially symmetric steady motion of a viscous incompressible fluid: some numerical experiments, *Computer J.*, *7*; 47.

Becker, H. A., Rosensweig, R. G., and Gwozdz, J. R. (1966). Turbulent dispersion in a pipe flow, *AIChE J.*, *12*: 964.

Bell, R. J. (1972). *Introductory Fourier Transform Spectroscopy*, Academic Press, New York.

Briggs, G. A. (1969). "Plume Rise," U.S.A.E.C., Division of Technical Information.

Bringfelt, B. (1968). Plume rise measurements at industrial chimneys, *Atmospheric Environment*, *2*: 575–598.

Carnahan, B., Luther, H. A., and Wilkes, J. O. (1969). *Applied Numerical Methods*, John Wiley & Sons, New York, pp. 440–442.

Corrsin, S., and Uberoi, M. S. (1950). N.A.C.A. Report No. 998.

Davis, D. D., Smith, G., and Klauber, G. (1974). Trace gas analysis of power plant plumes via aircraft measurement. Ozone, nitrogen oxides (NO_x), and sulfur dioxide chemistry, *Science, 186*: 732.

Fang, C. S., and Prengle, H. W., Jr. (1974). Radiation shape factors for remote gas volumes, *Trans. ASME, J. Heat Transfer, 49C*: 551.

Kleinstein, G. (1964). Mixing in turbulent axially symmetric free jets, *J. Spacecraft Rockets, 1*: 403.

Mahagaokar, U. (1976). Remote sensing of temperature and composition of gas plumes by infrared emission interferometry spectroscopy, Ph.D. Dissertation, Chemical Engr. Dept, Univ. of Houston, Houston, Texas.

Morton, B. R., Taylor, G. I., and Turner, J. S. (1956). Turbulent gravitational convection from maintained and instantaneous sources, *Proc. Roy. Soc., London, Ser. A., 234*: 1.

Perry, R. H., and Chilton, C. H. (1973). *Chemical Engineers' Handbook*, 5th edition, McGraw-Hill, New York, Section 5.

Prengle, H. W., Jr., Morgan, C. A., Fang, C. S., Huang, L. K., Campani, P., and Wu, W. W. (1973). Infrared remote sensing and determination of pollutants in gas plumes, *Environ. Sci. Tech. 7*: 417.

Priestly, C. H. B., and Ball, F. K. (1955). Continuous convection from an isolated source of heat, *Quart. J. Roy. Meteorol. Soc., 81*: 144.

Rosensweig, R. G., Hottel, H. C., and Williams, G. C. (1961). Smoke-scattered light measurement of turbulent concentration fluctuations, *Chem. Eng. Sc., 15*: 111.

Rouse, H., Yih, C. S., and Humphreys, H. W. (1952). Gravitational convection from a boundary source, *Tellus, 4*: 201.

Schmidt, W. (1941). Turbulent spreading of a stream of preheated air, *Z. Agnew. Math. Mech., 21*: 265.

Sutton, O. G. (1953). *Micrometeorology*, McGraw-Hill, New York, p. 10.

Tennessee Valley Authority (1954). Full Scale Study of Plume Rise at Large Electric Generating Stations, Muscle Shoals, Alabama.

Tomich, J. F., and Weger, Z. (1967). New results on momentum and heat transfer in compressible turbulent free jets, *AIChE J.*, *13*: 948.

3

Temperature Measurement by Infrared Spectrometry

LARRY A. GROSS and PETER R. GRIFFITHS *Department of Chemistry, University of California, Riverside, Riverside, California*

JEFFREY N.-P. SUN *Analect Instruments Division of Laser Precision Corp., Irvine, California*

1 INTRODUCTION

Over the past two decades, several methods of estimating the temperature of gases through measurement of their infrared spectra have been described. These techniques generally assume that the spectrum is measured at a sufficiently high resolution that the rotational fine structure may be distinguished, since the relative population of excited rotational energy states is strongly temperature dependent. Infrared spectrometric methods of estimating the temperature of gases may often be applied when the temperature may not be measured directly, e.g., by a thermocouple. As shown in Chap. 2, temperatures may be estimated from total in-band radiances. In this chapter, we will focus on spectral techniques, in which temperature accuracy depends on relative, not absolute, measurements. They also have the advantage that once the temperature is known, the concentration of each gaseous species with an infrared-active vibrational transition may be measured. Measurements of this type have become important for a variety of applications, including the modeling of planetary atmospheres (Rinsland et al., 1983), optimization of infrared gas lasers (Rolland et al., 1984), and remote sensing (Herget and Brasher, 1980).

In this chapter we will briefly summarize the instrumental techniques by which suitable infrared spectra may be measured. The theory of rotation–vibration spectra of simple molecules will be introduced, and the factors determining the shape of rotational lines and the temperature dependence of line profiles will be discussed. The effect of measuring the spectrum at a finite resolution will be introduced. Several methods for estimating gas temperatures spectroscopically will be described, and the results of some practical experiments reported.

2 INSTRUMENTAL CONSIDERATIONS

The measurement of infrared spectra of hot gases often involves detection of radiation emitted from very weak sources. Thermal emission from hot gases is rarely strong unless the temperature of the gas is very high. Even when absorption spectra are measured, it is rare that the radiation from a remote source can be efficiently collected, and thus the signal from the infrared detector is generally very low, especially if the spectrum is measured at high resolution. Bearing these considerations in mind, it is not surprising that the use of Fourier transform infrared (FT-IR) spectrometry in this field is becoming increasingly popular, due to several advantages to be noted below. It should be noted, however, that several other types of infrared spectrometers, including grating monochromators, Fabry–Perot interferometers, and tunable lasers, have also been used successfully for measurements of this type.

The principles of FT-IR spectrometry have been covered in a number of books and articles (Griffiths, 1975a, 1978, 1983). A detector measures the variation in intensity of the infrared beam emerging from a two-beam interferometer (such as a Michelson interferometer) as a function of the optical path difference between the two beams introduced by a moving optical element. This signal, termed an interferogram, is digitized and stored. The infrared spectrum is computed from the interferogram through the complex Fourier transform. In contrast to infrared spectrophotometers incorporating a grating monochromator, where each wavelength is sequentially passed across the exit slit of a monochromator, all wavelengths are detected simultaneously in an FT-IR spectrometer, giving rise to the multiplex (or Fellgett's) advantage. FT-IR spectrometers have a further advantage over scanning monochromators in that, for measurements made at a given resolution, their light-gathering power (or optical throughput) is also greater; this advantage is often known as Jacquinot's advantage.

For measurement of the infrared spectrum of weak sources, the combination of Fellgett's and Jacquinot's advantages can lead to an improvement in signal-to-noise ratio (SNR) of FT-IR spectra measured in a given measurement time of one to three orders of magnitude compared with a measurement made at the same resolution using a dispersive spectrometer. Because of its sensitivity for measuring the infrared spectra of weak sources, FT-IR spectrometry was first used for studying the infrared spectra of distant objects in space. Measurements of this nature have since become even more important with interferometers being mounted in spacecraft, on satellite platforms, and in high altitude balloons.

For terrestrial measurements, interference due to the absorption of infrared radiation by atmospheric water vapor and carbon dioxide can present severe problems. Nevertheless there are enough transmission "windows" to permit useful infrared spectra to be measured over fairly long paths through the atmosphere to cover at least half the mid-infrared spectral region (4000–400 cm^{-1}). Emission spectra of remote sources can often be measured from distances of up to 2 or 3 km. For absorption measurements even longer pathlengths can be tolerated, permitting sensitive detection of gases present in the atmosphere at concentrations a few parts per hundred million. A particularly important class of measurements falling into this category involves the use of the sun as a radiation source, especially at dawn or dusk when the path through the atmosphere can be greater than 100 km (Rinsland et al., 1983).

As early as 1967, Low and Clancy showed that the emission spectrum of the hot gases emerging from a smokestack could be measured using a low resolution rapid-scanning Michelson interferometer. The presence of SO_2 in the effluent gas could be readily detected from the emission spectrum. The resolution of this early interferometer (16 cm^{-1}) was so low, however, that no rotational fine structure could be resolved, and no attempt to determine the temperature of the gas or the concentration of SO_2 was made. More recently emission spectra have been measured at far higher resolution (Herget and Brasher, 1979). Examples of emission spectra of industrial stack gases measured by these authors are shown in Fig. 1. Emission bands of CO_2, CO, SO_2, and NO are readily seen in the spectra. Very weak lines due to HCl may also be observed. The fundamental band of CO_2 centered at 2347 cm^{-1} is greatly distorted because of absorption by atmospheric CO_2 at low temperature in the path between the flare and the detector. On the other hand, the spectra of the two diatomic molecules, CO and HCl, are relatively unaffected by atmospheric species. This observation is important, since the infrared spectra of diatomic molecules are more easily treated from a theoretical standpoint than are the spectra of polyatomic molecules, especially those of low symmetry. In Sec. 3 the theory of temperature measurements from emission or absorption spectra of diatomic molecules will be introduced. This theory can be extended to include polyatomic molecules, but for the purposes of this chapter we will only discuss the theory of the rotation–vibration spectra of diatomic molecules.

3 SPECTRA OF DIATOMIC MOLECULES

Emission and absorption spectra of gaseous samples are related by a very simple expression:

$$\varepsilon(\bar{\nu}) + \tau(\bar{\nu}) = 1 \tag{1}$$

where $\varepsilon(\bar{\nu})$ is the *emissivity* of the gas at wavenumber, $\bar{\nu}$ cm^{-1}, and $\tau(\bar{\nu})$ is its *transmittance* at $\bar{\nu}$. The emissivity of any sample is the ratio of its radiance compared to that of a blackbody at the same temperature. The radiance, $N_{\bar{\nu}}$, of any source over a small interval of the spectrum has units of energy per unit area per wavenumber. Thus $\varepsilon(\bar{\nu})$ is a dimensionless number which can be expressed as

$$\varepsilon(\bar{\nu}) = \frac{N_{\bar{\nu}}}{N^0_{\bar{\nu}}} \tag{2}$$

where $N^0_{\bar{\nu}}$ is the radiance of the blackbody at the same temperature. $\tau(\bar{\nu})$ is the ratio of the intensity of radiation incident on an absorbing sample at $\bar{\nu}$ to the intensity of radiation emerging from the sample at the same wavenumber. Therefore $\varepsilon(\bar{\nu})$ must always have a value between zero and unity.

Provided that the emissivity can be calculated from the measured emission spectrum, the transmittance (and hence the absorbance A) can be easily calculated, since

$$A(\bar{\nu}) = -\ln \tau(\bar{\nu}) = -\ln[1 - \varepsilon(\bar{\nu})] \tag{3}$$

For a pure gas, Lambert's law gives that

$$A(\bar{\nu}) = \kappa(\bar{\nu})l \tag{4}$$

where l is the path length (in meters), $\kappa(\bar{\nu})$ the absorption coefficient, and m^{-1} the unit of absorption coefficient at wavenumber $\bar{\nu}$. For lines in a rotation-vibration, or rovibrational spectrum, $\kappa(\bar{\nu})$ is strongly dependent on temperature. However it is very rare for any worker to attempt to measure $\kappa(\bar{\nu})$ for each line in the spectrum, since this would require a spectrometer with exceptionally high resolution. It is more common to determine the integrated absorption of a line, S, defined by

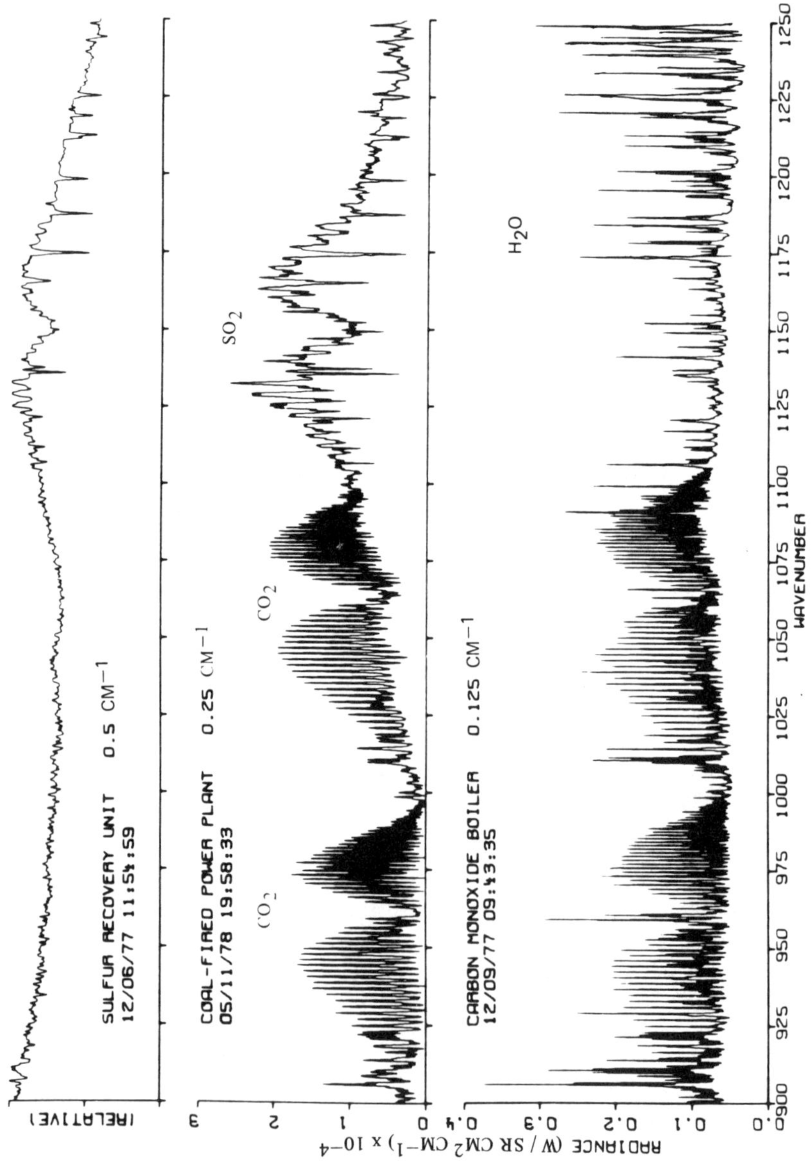
SULFUR RECOVERY UNIT 0.5 CM-1
12/06/77 11:54:59
(RELATIVE)
COAL-FIRED POWER PLANT 0.25 CM-1
05/11/78 19:58:33
CO_2
CO_2
SO_2
CARBON MONOXIDE BOILER 0.125 CM-1
12/09/77 09:43:35
H_2O
RADIANCE (W / SR CM2 CM-1) x 10-4
WAVENUMBER
900 925 950 975 1000 1025 1050 1075 1100 1125 1150 1175 1200 1225 1250
0 1 2 3
0.0 0.1 0.2 0.3 0.4

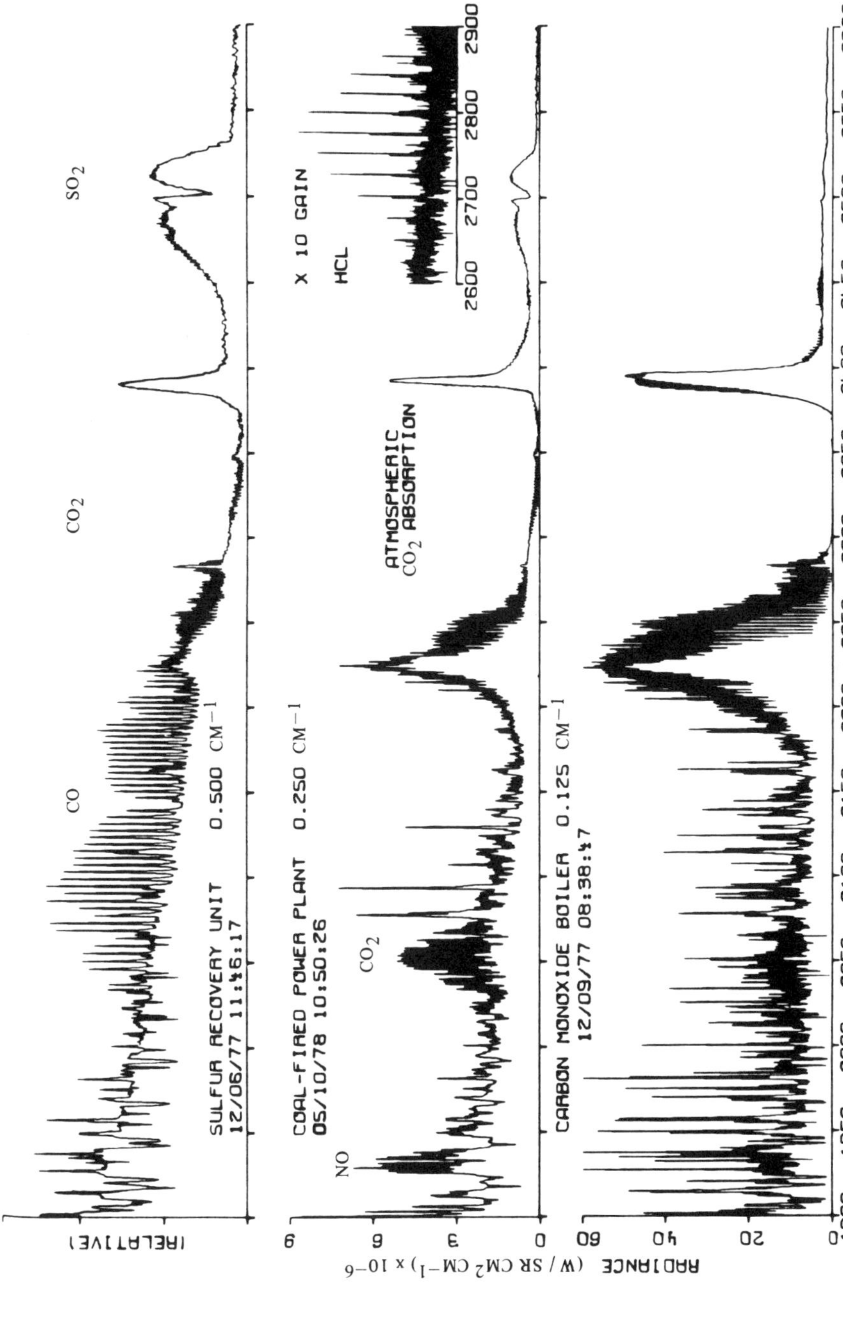

Fig. 1 Emission spectra of several industrial sources (from Herget and Brasher, 1979).

$$S = \int_0^\infty \kappa(\bar{\nu})\, d\bar{\nu} \tag{5}$$

since measurement of S can be made fairly accurately with instruments of moderate resolution.

Each line in the rovibrational spectrum of a diatomic molecule is due to a transition between two vibrational energy states, one of which is usually the ground state (characterized by a vibrational quantum number $\nu = 0$). At the same time a transition between two rotational states (each characterized by a rotational quantum number J) occurs. The selection rules for allowed transitions are

$$\Delta\nu = \pm 1$$
$$\Delta J = \pm 1 \tag{6}$$

where $\Delta\nu = \pm 1$ corresponds to the absorption of radiation and $\Delta\nu = -1$ corresponds to emission. (Overtone bands, for which $\Delta\nu$ is greater than 1, are typically much weaker.) Transitions in which $\Delta J = +1$ occur at higher energy than the band center, and the lines in the rotation-vibration spectrum corresponding to these transitions are referred to as the R branch of the spectrum. Similarly, transitions in which $\Delta J = -1$ occur at lower energy than the band center and comprise the P branch of the spectrum.

Each line in the P branch is characterized by a symbol P(J''), where J'' is the rotational quantum number of the lower state. Since $\Delta J = -1$ for this branch, the first line in the P branch is P(1). Each line in the R branch is characterized by the symbol R(J''). Since $\Delta J = +1$ for the R branch, the first line in this branch is R(0). The line strengths in each branch go through a maximum value, which increases with the temperature of the gas. An emission spectrum of CO at 473 K is shown in Fig. 2. It can be seen from this spectrum that the maxima in the P and R branches occur at approximately P(10) and R(10) and that lines beyond P(30) and R(30) may be observed. As the temperature of the gas is increased, the higher rotational energy levels become more highly populated. By comparing the absorption or emission of each line in the spectrum with a theoretically calculated spectrum, it should be possible to estimate the tempera-

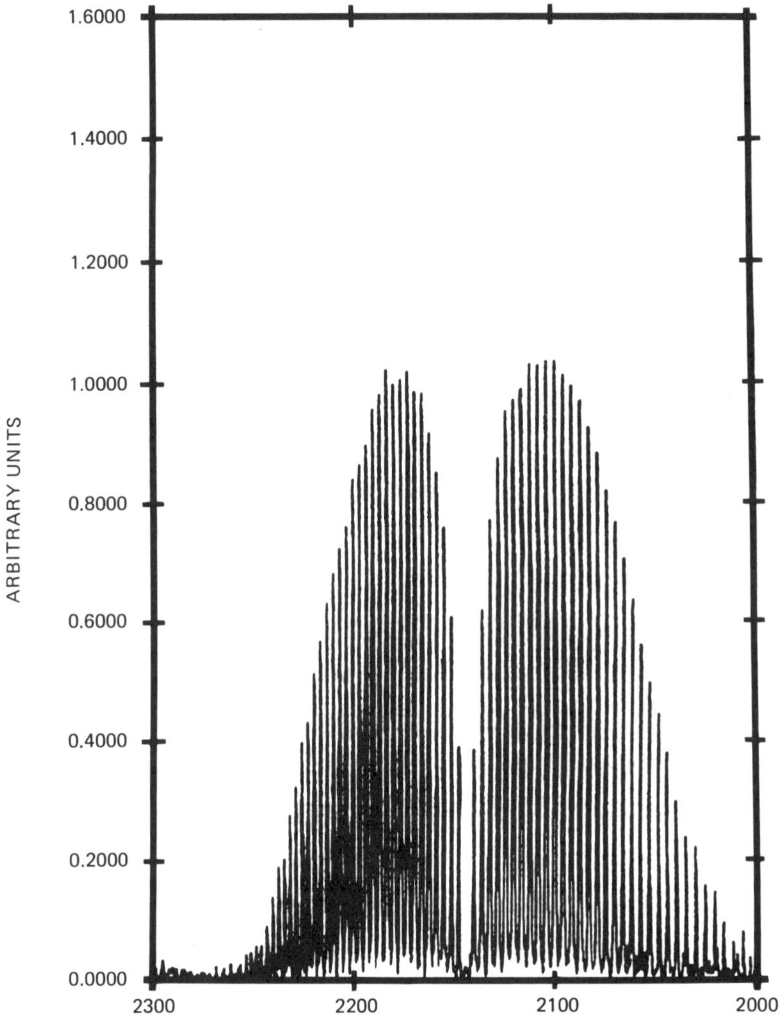

Fig. 2 Emission spectrum of 1% CO in N_2 at 473 K and a total pressure of 760 Torr; the cell pathlength was 10 cm.

ture of the gas, although in practice this is usually not as trivial as it may sound due to a variety of factors, some of which will be discussed later in this chapter.

Since the width of an absorption line depends on the pressure of the gas, $\kappa(\bar{\nu})$ is usually not directly proportional to concentration. However, for a gas at a sufficiently low density (optically thin), the integrated absorption S is directly proportional to the number of molecules of the absorber in the optical path. S may be set equal to the product of the line strength, S_0, and the pressure P of the absorber. The sum of the strengths of the individual lines in the entire band is defined as the band intensity S_0:

$$S_0 = \sum_m S_0(m) \tag{7}$$

Here m is the *index* of a particular line and equals $J'' + 1$ for an R branch line and $-J''$ for a P branch line, where J'' is the rotational quantum number of the initial state and $S_0(m)$ is the line strength of a line of index m. The sum is taken over all the lines in the band.

The line strength as defined above is directly related to the quantum mechanical probability of a transition. For the fundamental band of CO, the total band strength S_0 has the value of 284 $\text{cm}^{-2}\ \text{atm}^{-1}$. A considerable research effort has gone into establishing this value accurately because it is the spectral parameter that relates the intensity of the CO band in the spectrum to the amount, or partial pressure, of CO in the sample. Hanson (1983) has discussed how measurements of spectral line strengths for CO have been used in his laboratory as an aid for identifying the origin of a particular line. From the measurement of the line strength of a single line, the line strength $S_0(m)$ for all other lines in that band can be calculated if the partition function for the band is known and if the normal intensity distribution is unperturbed. (Vibration-rotation interaction would represent such a perturbation.) The partition functions for diatomic molecules have been discussed by Herzberg (1950). Several useful partition functions have been tabulated in a recent article (Chackerian et al., 1983).

The intensities of individual lines are proportional to the population of molecules in the particular energy level from which the transition originates. The distribution of population among the rotational energy levels follows the Maxwell–Boltzmann law:

$$N_{J''} = N_0 \frac{(2J'' + 1)}{Q_{rot}} \exp\left[\frac{-E(J'')}{kT}\right] \tag{8}$$

where N_0 and $N_{J''}$ are the total number of molecules in the initial vibrational state and the number in the J'' rotational level, respectively; Q_{rot} is the rotational partition function, k the Boltzmann constant, and T the absolute temperature. The rotational energy of the initial state of the transition is given to a good approximation by

$$E(J'') = hc[B''J''(J'' + 1) - D''J''^2(J'' + 1)^2] \tag{9}$$

where B'' is the so-called rotational constant, and is related to the atomic weights of the two atoms and the distance between them, and D'' the centrifugal distortion constant. Spectral line frequencies are proportional to the differences of energies given by Eq. (9) for the upper state. The frequency of an individual line with index m is given by (Herzberg, 1950)

$$\begin{aligned}\bar{\nu} = \bar{\nu}_0 &+ (B' + B'')m + (B' - B'' - D' + D'')m^2 \\ &-2(D' + D'')m^3 - (D' - D'')m^4\end{aligned} \tag{10}$$

The band center is at $\bar{\nu}_0$, and B' and D' are the values of the constants defined above for the upper vibrational state. Typically B is much greater than D, and the B values for the upper and lower states are similar. When this is true, it is easily seen from the selection rules given in Eq. (6) that the lines in the rovibrational spectrum of a diatomic molecule are approximately equally spaced. The effect of D is to decrease the spacing of lines in the R branch as J'' increases and to increase the spacing of lines in the P branch.

For diatomic molecules in thermal equilibrium with their surroundings, the intensity of a rovibrational line is determined by the population expression [Eq. (8)]. However, the transition probability is not precisely the same for all lines in the band, and this is taken into account by using the mean value of $(2J + 1)$ for the upper and lower states. Therefore $(2J'' + 1)$ is replaced by $(J' + J'' + 1)$ in the expression for the integrated absorption $S(m)$ for each line

$$S(m) \propto (J' + J'' + 1)F(m)\bar{\nu}(m) \exp[-E(J'')/kT) \tag{11}$$

where J' is the rotational quantum number for the upper state and $F(m)$ the so-called *Herman–Wallis factor*. $F(m)$ accounts for the effect of rovibrational interaction on intensities. It is usually represented empirically in the form

$$F(m) = 1 + c_1 m + c_2 m^2 \tag{12}$$

For the CO molecule, the values of c_1 and c_2 are 1.77×10^{-4} and 6.67×10^{-6}, respectively, so the effect of $F(m)$ on $S(m)$ is small but not negligible (Herman and Wallis, 1955). For very polar molecules such as HCl, c_1 and c_2 are significantly larger. For all but the lowest values of J, $(J' + J'' + 1)$ can be replaced in the proportionality shown in Eq. (11) by $|m|$, and this approximation is made throughout the rest of this chapter. In addition, since all parameters will subsequently be given in terms of m rather than J'', we denote $E(J'')$ as $E(m)$.

A plot of $\ln[S(m)/|m|F(m)\bar{\nu}(m)]$ against $E(m)$ should produce a straight line of slope $-1/kT$, from which the temperature could be obtained. However, $S(m)$ is a difficult quantity to measure due to baseline errors and the effect of overlapping bands, especially those due to different isotopic species such as ^{13}CO in the case of carbon monoxide and $H^{37}Cl$ in the case of hydrogen chloride. It is more desirable to be able to measure the peak height of each line, but a treatment involving peak heights is made more difficult because $S(m)$ depends both on the peak height and width.

If we consider an absorption spectrum of a diatomic molecule for which the peak shape is assumed to be Lorentzian, the absorbance at any wavenumber, $A(\bar{\nu})$, is related to the peak absorbance A_{peak}, occurring at wavenumber $\bar{\nu}_0$ by the expression

$$A(\bar{\nu}) = A_{peak} \frac{\gamma^2}{\gamma^2 + (\bar{\nu} - \bar{\nu}_0)^2} \tag{13}$$

where γ is the half-width of the band at half peak absorbance. (Note: γ is one half of the full width at half height, FWHH.) For a Lorentzian peak shape, $S(m)$ and $A_{peak}(m)$ are related by

$$S(m) = \pi A_{peak}(m)\gamma(m) \tag{14}$$

Thus Eq. (11) becomes

$$A_{peak}(m) \propto \frac{|m|F(m)\bar{\nu}(m)\exp[-E(m)/kT]}{\gamma(m)} \tag{15}$$

In this case, $\ln[A_{peak}(m)\gamma(m)/|m|F(m)\bar{\nu}(m)]$ should be plotted against $E(m)$ to obtain the temperature of the gas.

It is important to recognize that the use of peak absorbances requires a knowledge of the widths of the individual lines of a vibrational–rotational band, since the widths vary with m. It is even more necessary to recognize when measuring these very narrow lines that photometric errors can be introduced by convolution of the true line profile with the instrument line shape (ILS) function of the spectrometer which was used for the measurement. Since the widths of individual lines in the spectrum of self-broadened or air-broadened carbon monoxide are typically on the order of 0.1 cm^{-1}, all but the very highest resolution instruments will seriously distort the line shape and will lead to a measured, or apparent, absorbance at the peak of a line, A^a_{peak}, which is much less than the true peak absorbance A^t_{peak}. Only the values of A^t_{peak}, not A^a_{peak}, can be used in Eq. (15).

4 RETRIEVING SPECTROSCOPIC ESTIMATES OF TEMPERATURE

4.1 Band-Fitting Methods

Anderson and Griffiths (1977) have developed a method for temperature measurements that involves a nonlinear least squares fit of calculated values of A_{peak} to the peak absorbance of the carbon monoxide rovibrational lines. The true peak absorbance can be calculated by rewriting Eq. (15) as

$$A^t_{peak}(m) = \frac{A_1|m|F(m)\bar{\nu}(m)\gamma(1)\exp[-E(m)/kT]}{F(1)\bar{\nu}(1)\gamma(m)} \tag{16}$$

where A_1 is the true peak absorbance of the $m = 1$ line. The method begins with a good guess of the temperature T of the sample, and A_1 can be *approximated* by the measured absorbance of the first line in the R branch, corrected for the photometric error introduced by the finite instrument resolution. The steps outlined below are then followed to calculate a change in temperature, ΔT, and change in A_1, ΔA_1, which are added to T and A_1 to give a new estimate for these two parameters. $T + \Delta T$, and $A_1 + \Delta A_1$, the new estimates, are substituted for T and A_1 in Eq. (16), and new values are obtained for ΔT and ΔA_1. This procedure of successive approximation continues until the difference between the calculated absorbance and the experimentally measured absorbance of each peak becomes small. This difference is termed a residual, and at each step T and A_1 are adjusted to minimize the sum of the squares of the residual absorbances for the peaks.

The matrix algebra for the procedure used is described by Kim (1970) and Bard (1974). This is the familiar method of treating the sum of the squares of the residuals as a function of A_1 and T, and holding the residuals of the individual peaks constant. The derivative of this matrix function is set equal to zero in order to find its minimum, and the matrix of coefficients is inverted in order to solve for the values of ΔA_1 and ΔT. Two related requirements must be satisfied for this method to work. First, A_1 and T must be independent of each other if a solution for ΔA_1 and ΔT is to be obtained. In other words, two equations in two unknowns cannot be solved unless the equations are independent. Second, the initial estimates of A_1 and T must be fairly good to ensure that the successive approximations will indeed converge to values that will give a best fit for the experimental spectrum.

The calculation of the residuals needed for the above procedure occurs in three steps. First, the nonlinear exponential dependence of the true peak absorbance $A^t_{peak}(m)$ is calculated for each carbon monoxide line of rotational index m. The second step is the calculation of $A^a_{peak}(m)$, the apparent peak absorbance, from $A^t_{peak}(m)$. The method of this calculation is given below, and it is discussed in more detail in another paper by Anderson and Griffiths (1975).

$A^a_{peak}(m)$ is the convolution of the theoretical value of $A^t_{peak}(m)$ for a given line at a given temperature and the ILS function. After $A^a_{peak}(m)$ has been calculated, it is compared with the measured absorbance recorded by the spectrometer (the third step) by subtracting each calculated value of $A^a_{peak}(m)$ from the measured peak absorbances. Thus the residuals needed for the matrix algebra outlined above are obtained as a vector with an element corresponding to each value of the line index m.

The potential sensitivity of this method for measuring temperature is limited by the errors introduced at each of the above steps. For instance the linewidth $\gamma(m)$ depends on temperature T and pressure p in both the calculation of $A^a_{peak}(m)$ and $A^t_{peak}(m)$. Laboratory studies of the dependence of $\gamma(m)$ on T and p have been initiated to reduce the uncertainty in this parameter. This work will be discussed in some detail later in this chapter. Values of $\gamma(m)$ are needed for each of the rotational lines for Eq. (16), and these can be calculated from known values of $\gamma^0(m)$, the line half-width at half maximum at standard temperature (273 K) and pressure (760 Torr). Studies of the variation of $\gamma(m)$ with temperature and pressure have been made using a very high resolution spectrometer to establish this functional dependence, with the goal of improving the accuracy of the values calculated for $A_{peak}(m)$. In comparison to the other factors in the equation for $A_{peak}(m)$, $\gamma(m)$ is known rather imprecisely, so the degree of improvement possible and the effect on the accuracy of the temperature-sensing method described will also be of interest. The utmost importance of accurate values of $\gamma(m)$ for the calculation of $A^t_{peak}(m)$, the second step above, will become clear in the following discussion.

It has already been noted that the reason for using the absorbance at the peak maximum A_{peak} in the calculation to retrieve temperature was to reduce the error in the experimentally recorded integrated absorption $S(m)$. When emission spectra instead of absorption spectra are being measured, the complementary relationship between emissivity and transmittance in Eq. (1) must be used to compare the theoretical intensities with the measured data. This relation is based on the assumption that the gases whose spectra are being measured are at equilibrium, which may not be the case, and further studies will have to be made to examine this problem. For this reason and because the intensity of the measured emission spectrum is dependent on geometric factors such as the distance to the source and the size of the collecting telescope, it is more practical to develop a theoretical expression for the apparent peak transmittance $\tau^a_{peak}(m)$, which can then be compared with the measured emission spectrum multiplied by a function that corresponds to the instrument response at different wavelengths.

The resolution of a spectrometer is quantitatively described by its instrument line shape (ILS) function $g(\bar{\nu} - \bar{\nu}_i)$, which describes the response of the instrument to radiation of wavenumber $\bar{\nu}$ when the instrument measures at a wavenumber of $\bar{\nu}_i$. If the radiant power at wavenumber $\bar{\nu}$ incident on the spectrometer is given by $P(\bar{\nu})$, the apparent intensity recorded by the spectrometer at wavenumber $\bar{\nu}_i$ is given by the convolution of $P(\bar{\nu})$ with the ILS function:

$$P^a(\bar{\nu}_i) = \int_0^\infty g(\bar{\nu} - \bar{\nu}_i)P(\bar{\nu})\, d\bar{\nu} \tag{17}$$

For absorption measurements, the radiant power incident on the sample, $P_0(\bar{\nu})$, which is transmitted to the detector is

$$P(\bar{\nu}) = P_0(\bar{\nu}) \exp(-\kappa(\bar{\nu})l) \tag{18}$$

If $P_0(\bar{\nu})$ is taken to be constant over the spectral bandpass of the instrument, the apparent transmittance at wavenumber $\bar{\nu}_i$ becomes

$$\tau^a_{peak}(\bar{\nu}_i) = \frac{\int_0^\infty g(\bar{\nu} - \bar{\nu}_i) \exp(-\kappa(\bar{\nu})l)\, d\bar{\nu}}{\int_0^\infty g(\bar{\nu} - \bar{\nu}_i)\, d\bar{\nu}} \tag{19}$$

For these calculations, the absorption feature is assumed to have a Lorentzian shape [see Eq. (13)]. Such a contour is expected for a single absorption line for a molecule in the gas phase whose shape is determined by collision broadening.

The ILS of an FT-IR spectrometer depends on the way in which the interferogram is apodized prior to the Fourier transform (Griffiths, 1975b). If the interferogram is unapodized, the ILS is a sinc [(sin x)/x] function. When triangular apodization is used, the ILS is a sinc^2 function:

$$g(\bar{\nu} - \bar{\nu}_i) = \sin^2(\pi\delta_0(\bar{\nu} - \bar{\nu}_i))/(\pi\delta_0(\bar{\nu} - \bar{\nu}_i))^2 \tag{20}$$

where δ_0 represents the maximum optical retardation used in collecting the interferogram.

It was also found useful to define a resolution parameter ρ, which specifies the resolution of the spectrometer as compared to the FWHH of the line being measured (Anderson and Griffiths, 1975):

$$\rho = \frac{\mathsf{R}}{\text{FWHH}} = \frac{1}{2\gamma\delta_0} \tag{21}$$

The resolution R has been represented by $1/\delta_0$, since this is the value that is conventionally taken as the nominal resolution inherent in the interferogram. For the sinc^2 function, $1/\delta_0$ is the distance from the center of the bandpass to the first zero in the ILS function. This is the Rayleigh criterion, that two resonances be considered resolved if the peak of one falls at the first zero of the second.

Substitution of Eqs. (13), (20), and (21) into Eq. (19) allows computation of the apparent line shapes by means of successive numerical integrations for selected values of $\bar{\nu}_i$ on the curve. Normally the complete band profile is of little interest, whereas the apparent peak absorbance $A^a_{peak}(m)$ is very important. By setting $\bar{\nu}_i = \bar{\nu}_0$ in Eq. (19) and defining

$$y = (\bar{\nu} - \bar{\nu}_i)/R \tag{22}$$

$$g(y) = 2\sin^2(\pi y)/(\pi y)^2 \tag{23}$$

the expression for the apparent minimum transmittance becomes

$$\tau^a_{peak} = \int_0^\infty g(y)\exp\left(\frac{-A^t_{peak}}{1 + 4\rho^2 y^2}\right) dy \tag{24}$$

Correcting the resolution error by substituting Eq. (16) into Eq. (24), $\tau^a_{peak}(m)$ can be calculated as a function of $A^t_{peak}(m)$ and $\boldsymbol{\rho}$. The resolution parameter $\boldsymbol{\rho}$, for a known sample pressure, is a function of temperature through the temperature dependence of γ, which will be discussed later. Thus $\tau^a_{peak}(m)$ is a function only of A_1 and T and can be calculated from those parameters by way of Eqs. (16) and (24). The approach then is to adjust A_1 and T by using a nonlinear least squares technique, so that the calculated peak transmittances best reproduce the experimental ones. A computer program (called ROTMP) was written to best reproduce the experimental transmittance.

A modification of this program, called ROTMPE, was also written to permit temperatures to be estimated from the emission spectra of diatomic molecules. For emission spectra the height of the blackbody curve is equivalent to $P^0(\bar{\nu})$ for transmission spectra. This is completely unknown, but if the blackbody intensity is known at one wavenumber, it can be calculated across the entire spectrum. Thus a third parameter, the height of the blackbody curve at the wavenumber of the R(0) line, was included in ROTMPE.

4.2 Graphical Methods

From Eq. (10) it can be seen that a plot of $\ln[S(m)/|m|F(m)\bar{\nu}(m)]$ against $E(m)$ should produce a straight line with a slope of $-1/kT$. For absorption spectra, $S(m)$ is equal to $\pi A_{peak}(m)\gamma(m)$, so that a plot of $\ln[A_{peak}(m)\gamma(m)/|m|E(m)\bar{\nu}(m)]$ against $E(m)$ should be linear and yield the temperature from its slope. In many cases, the linewidths are less than the FWHH of the ILS function of the spectrometer used to measure the spectrum, i.e., the resolution parameter $\boldsymbol{\rho}$, defined in Eq. (21), is much greater than unity.

As we noted above, the effect of the finite resolution is to reduce A^t_{peak} to a much smaller value, A^a_{peak}. If the true peak absorbance and $\boldsymbol{\rho}$ are both large, it can be shown that A^a_{peak} is approximately proportional to the square root of the true value of the peak absorbance A^t_{peak}. For spectra measured on a Fourier spectrometer with triangular apodization, Anderson and Griffiths (1975) showed that

$$A^a_{peak} \cong 1.1(A^t_{peak})^{1/2}/\boldsymbol{\rho} \tag{25}$$

A similar relationship applies to grating spectrometers (Nielsen et al., 1944). Under these conditions,

$$(A^a_{peak}(m))^2 \propto |m|F(m)\bar{\nu}(m)\gamma(m)\exp(-E(m)/kT) \tag{26}$$

and a plot of $\ln[(A^a_{peak}(m))^2/|m|F(m)\bar{\nu}(m)\gamma(m)]$ against $E(m)$ will be linear.

For emission spectra, $S(m)$ varies with the fourth power of wavenumber as a result of the Einstein coefficient for spontaneous emission. In this case,

$$S(m) \propto |m|F(m)[\bar{\nu}(m)]^4\exp(-E(m)/kT) \tag{27}$$

The measured radiance $N(m)$ at a line center will be proportional to the product of the emissivity at the wavenumber, $\varepsilon(m)$, and the radiance from a blackbody at that temperature, N^0. Therefore by application of Eq. (1):

$$N(m) \propto (1 - \tau(m))N^0 \tag{28}$$

For any spectral line:

$$1 - \tau = A_{peak} - \frac{A_{peak}^2}{2!} + \frac{A_{peak}^3}{3!} \cdots \tag{29}$$

Thus for a weak line, $N(m)$ will be approximately proportional to the product of $A_{peak}(m)$ and N^0. If it is assumed that N^0 is constant across a spectral band, a plot of $\ln[N(m)\gamma(m)/|m|F(m)[\bar{\nu}(m)]^4]$ against $E(m)$ will be linear, provided that resolution errors are insignificant.

As for band-fitting methods, it must be noted that graphical methods applied to absorption or emission spectra require good values for the linewidths. In particular, an accurate knowledge of the temperature dependence of pressure broadening coefficients is needed.

4.3 Two-Line Method

Although resolution errors are commonly encountered when gas phase spectra are measured by an FT-IR spectrometer, much higher resolution may be routinely attained when spectra are measured using a tunable diode laser (TDL) spectrometer. The diode itself is a semiconductor less than 1 mm^2 in dimension, consisting of a ternary lead salt crystal typically having a composition such as $PbS_{1-x}Se_x$. The desired wavelength range is selected during the manufacture of the diode by the choice of the doping elements and by controlling the fraction, given by x, of their amounts in the crystal. When the diode is cooled to cryogenic temperatures and forward-biased with a current of 2A or less, it emits several narrow lines of coherent infrared radiation which are each separated by about one wavenumber, due to different modes. Tuning over a narrow range (typically 1 cm^{-1}) is achieved by stabilizing the diode temperature variation to less than about 10^{-3} K and then varying the current. The maximum tuning range allowed by changing the temperature is anywhere from 20 to 100 cm^{-1}, depending on the manufacturing process by which the laser is constructed. Heterostructure laser diodes are now being developed to operate at temperatures up to 120 K, which gives them a range of accessible wavelengths significantly greater than the range of contemporary lead salt diodes (Partin, 1983). Eng and Ku (1982) have reviewed the theory of operation of TDL spectrometers and several applications to molecular spectroscopy.

It was mentioned earlier that isotopic lines would sometimes create an interference to the accurate measurement of temperature. Another source of interference when spectra are only measured at medium resolution is due to the presence of "hot bands", i.e., bands associated with transitions between two excited vibrational states. An example of this type of interference is shown in Fig. 3, where the line interfering with the P(17) line of ^{13}CO can easily be identified as a hot band because it is only observed at the higher temperature. The intensity of the interfering line is proportional to the population of molecules in the excited energy state from which the transition originated. This population is sensitive to temperature through the Boltzmann distribution in Eq. (8). For CO at the lower temperature in Fig. 2, this initial excited state is not appreciably populated and thus the line is not observed.

The appearance of hot bands at temperatures greater than ambient presents an alternative method for measuring temperature known as the two-line method. The variation in the intensity of

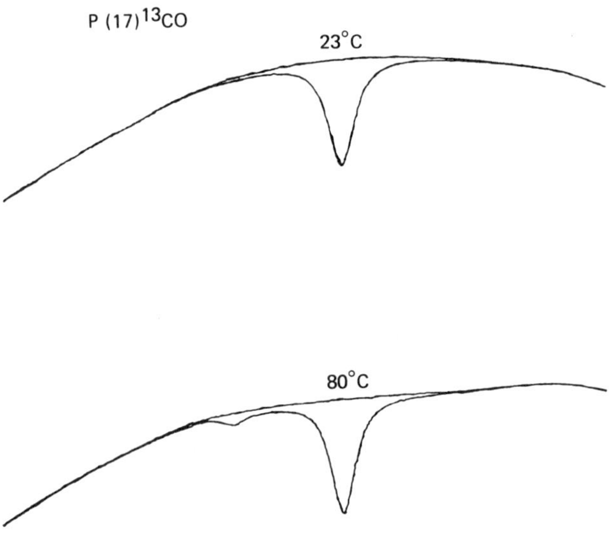

Fig. 3 Single-beam spectra of the P(17) line of $^{13}C^{16}O$ at 296 K (upper trace) and 353 K (lower trace). The spectrum of the empty cell is also shown for each trace. In the second spectrum, an interfering hot band can also be observed.

the hot band with temperature may be calculated by an equation similar to the one used to model the intensities of the rotational lines within a band. The two-line method is thus a similar spectroscopic probe and is applied to only two neighboring lines in different, but overlapping bands instead of all the lines in a single band (Penner, 1959). The availability of high resolution TDL spectrometers in recent years has made it possible to apply this method to lines in a rovibrational band, because a high resolution instrument is needed to record the line profile without distortion. When operating on a single mode, the tuning range of a TDL spectrometer is limited to about 1 cm^{-1}. It is apparent that the two lines should be within this range, but far enough apart to be completely resolved. Since the population of excited vibrational states is generally rather low, an intense line in the hot band should be ratioed against a weak line in the fundamental.

The two-line method also has in common with the FT-IR methods described in the previous sections the requirement for accurate knowledge of the half-width of a line and its variation with temperature. Hanson and Falcone (1978) have stated that the ratio of the logarithms of the minimum transmittance for each line is sometimes an acceptable approximation to the ratio of line intensities, but they stress that the two lines should have similar half-widths at the temperatures of interest. The same article lists a number of line pairs for which a line of a CO hot band is adjacent to a line of the fundamental band of CO, and gives the quantum numbers of each.

Wang (1976), like the authors above, has discussed the sensitivity of the experimental intensity ratio to temperature, which is defined as the slope of the plot of the intensity ratio versus temperature. A large sensitivity is desired for improved accuracy of the two-line temperature probe. Letting F stand for the ratio of intensities between the two lines, he derived the equation

$$\frac{\Delta F}{F} = \left(\frac{1}{T} + \frac{1.439\ \Delta E(m)}{T^2} \right) \Delta T \tag{30}$$

where T is absolute temperature and $\Delta E(m)$ is the difference in lower-state energies for the two rotational lines, the energy of each being given in terms of Eq. (9). From this treatment it was shown that the greatest sensitivity in the two-line method can be obtained by selecting lines which differ widely in their index m.

5 LINE BROADENING MECHANISMS

5.1 Natural Line Broadening

In the absence of all other broadening mechanisms, such as Doppler or collision broadening, the linewidth is governed by the Heisenberg uncertainty principle, i.e., the width is inversely proportional to the lifetime of the excited state. In the infrared region this lifetime is relatively long, about 10^{-5} to 10^{-4} sec. The natural width may be calculated from

$$\gamma_N = (2\pi c \Delta t)^{-1} \text{ cm}^{-1} \tag{31}$$

where c is the velocity of light (m sec^{-1}) and Δt is the lifetime of the excited state. The width of rotational lines solely due to natural broadening is therefore on the order of 10^{-7} cm^{-1}. In practice, the width of the lines is never determined by natural broadening, but rather by Doppler or collision broadening, or a combination of these two mechanisms.

5.2 Doppler Broadening

The Doppler effect is familiar as the shift in frequency that occurs due to the velocity of the molecule with respect to the radiation field. The direction of motion of individual molecules in an isotropic gas is random, so that Doppler-broadened lines have a Gaussian profile given by

$$\kappa_D(\bar{\nu}) = \left(\frac{\ln 2}{\pi}\right)^{1/2} \frac{S}{\gamma_D} \exp\left[\frac{-\ln 2(\bar{\nu} - \bar{\nu}_0)}{\gamma_D}\right]^2 \tag{32}$$

where S is the line intensity and γ_D is the Doppler-broadened half-width. This Gaussian shape is observed only in the spectra of gases at low pressure ($P < 100$ Torr for nonpolar molecules, and much less than this for polar molecules or for nonpolar molecules in the presence of polar molecules). The Doppler width is independent of pressure, but varies with the molecular weight of the molecule and the temperature. This dependence should be expected, since the Doppler effect is a consequence of the distribution of molecular velocities. Using a statistical mechanical argument, the Doppler width is found to be proportional to the average molecular velocity and is given by

$$\gamma_D = \left[\frac{2 \ln 2}{(mc^2/kT)}\right]^{1/2} = 3.58 \times 10^{-7}\bar{\nu}\left(\frac{T}{m}\right)^{1/2} \tag{33}$$

where m is the molecular weight. At 350 K, γ_D for CO (for which $\bar{\nu}$ = 2145 cm^{-1}) is 0.0027 cm^{-1}. Since γ_D is invariably larger than γ_N, Doppler broadening determines the line profile at low pressure.

5.3 Collision Broadening

When gases are present at pressures above a certain limit, collision broadening determines the line profile. For collision-broadened lines, the width γ_c is always linearly proportional to the pressure of the gas, so that for self-broadened gases:

$$\gamma_c = \gamma_0 p \tag{34}$$

where p is the pressure and γ_0 is the half-width at a reference pressure p_0. For a p_0 of 1 atm, p is the pressure in atmospheres, and the pressure-broadening coefficient γ_0 has units of cm^{-1} atm^{-1}. When an absorbing (active) gas is the only one present, γ_0 is known as the self-broadening coefficient. For a mixture of gases the width of a line in the spectrum of the active gas resulting from broadening by other gases is given by

$$\gamma_c = \sum_i \gamma_{0i} p_i \tag{35}$$

where γ_{0i} is the pressure-broadening coefficient of the absorbing gas under 1 atm of the ith component of the mixture, having units of cm^{-1} atm^{-1}. Since S is proportional to the number of molecules of the absorbing gas in the optical path, it is only affected by changes in the partial pressure of the active gas, whereas the half-width is determined by the total pressure.

The profile of collision-broadened lines should be Lorentzian, given by

$$\kappa_c(\bar{\nu}) = \frac{S}{\pi} \frac{\gamma_c}{(\bar{\nu} - \bar{\nu}_0)^2 + \gamma_c^2} \tag{36}$$

This functionality is derived using the interruption approximation, in which the duration of a collision is assumed to be negligibly short compared to the time required for a vibrational transition (Breene, 1981). The collisions are assumed to be elastic, i.e., there is no exchange of energy between the colliding molecules. A molecule which undergoes such a collision while in the process of a vibrational transition changes its orientation instantaneously, interrupting the transition. After the collision the transition continues, with the frequency of the transition remaining constant because the initial and final energy states are unaffected by the impact. The impact theory explains the pressure dependence of γ_c in terms of instantaneous random changes in the phase of the electromagnetic wave at the transition frequency rather than a change in the frequency. In addition, this theory predicts that the pressure-broadened half-width γ_c should be inversely proportional to the square root of temperature.

During a collision intermolecular forces vary with the distance R between the colliding molecules. The potential energy $V(R)$ of the long-range part of the interaction can be expressed as a function of distance R by the intermolecular potential $V(R) = C/R^n$, where the constant C determines the magnitude of the potential and whether it is attractive or repulsive (Hirschfelder et al., 1954; Maitland et al., 1981).

Anderson (1949) proposed a more general quantum mechanical treatment for collision-broadened spectral lines which took into account the details of the intermolecular potential. Anderson's theory of pressure broadening has been discussed in considerable detail and extended by Tsao and Curnutte (1962). By using a simplified version of the Anderson–Tsao–Curnutte theory, which plays the central role in line broadening studies, Birnbaum (1967) derived an equation for the temperature dependence of the linewidth:

$$\gamma \propto T^{-(q+4)/2q} = T^{-(n+1)/2(n-1)} \equiv T^{-N} \tag{37}$$

where q is a parameter given by Birnbaum which depends on the type of interaction between two molecules when they collide. The same formula can be derived from the classical line-broadening formulation (Hindmarsh and Farr, 1973). It is found that the temperature exponent N can be expressed in terms of the intermolecular potential exponent n, as seen in the second expression in Eq. (37).

A general theory of line broadening in the infrared would be valuable because it would alleviate the need for empirical linewidth measurements in problems concerning radiative transfer. The recent article by Bonamy et al. (1984) addresses this problem for CO, and their model successfully predicts the linewidth and temperature broadening exponent at elevated temperatures. This theory is not intended to apply for temperatures from ambient to about 600 K, because over this range the linewidths vary with the rotational quantum number J to a greater extent than they do at elevated temperatures. To simplify the calculations, average values for J were assumed and the results were improved by using a more realistic treatment of the collision process.

The forces arising from collisions depend on the positions of the electrons and nuclei of each molecule, whose distribution can be described by the summation of their dipole, quadrupole, and octupole moments. A list of the possible interactions and the appropriate values of q and n is given in Table 1. There have been few experimental measurements to determine the value of N in Eq. (37) until recently, due to the difficulty of the measurements. The quadrupole–quadrupole interaction, for which $N = 0.75$, has long been considered to be the most significant interaction in regard to collisional broadening of CO by CO itself and by N_2, and in fact studies of the temperature dependence of linewidth represent one method of measuring quadrupole moments for a molecule (Birnbaum, 1967).

The values of n, the exponent in the intermolecular potential function, are listed in Table 1 to indicate the range over which each of these forces are important. The exponent n is small for the strong, longer range forces, for example dipole–dipole interactions for which $V(R)$ is large when R is large. Dipole–dipole forces are important for interactions between molecules with a large dipole moment, like HCl. Larger values for the exponent n cause $V(R)$ to become relatively small when R is large and indicate a force which is shorter in range. Finding experimental values for N smaller than 0.75 does not mean that quadrupole–quadrupole interactions are not dominant, as they are for CO collisions, but instead it means that shorter range forces are also important and should not be neglected in theoretical calculations.

The dependence of the collision-broadened linewidth on temperature and pressure is usually given as

$$\gamma_c = \gamma^0 \left(\frac{p}{p_0}\right)\left(\frac{T_0}{T}\right)^N \tag{38}$$

TABLE 1 Parameters q, N, and n for Various Intermolecular Potentials

Type of interaction	q	N	n
Dipole–dipole	4	1	3
Dipole–quadrupole	6	5/6	4
Quadrupole–dipole	6	5/6	4
Quadrupole–quadrupole	8	3/4	5
Dipole-induced dipole	10	7/10	6
Dispersion	10	7/10	6
Quadrupole-induced dipole	12	2/3	7

where p_0 and T_0 refer to the standard conditions of pressure and temperature, 1 atm and 297 K, respectively. The temperature dependence of the R(0), R(8), and R(16) lines in the fundamental band of CO under conditions of nitrogen broadening was studied by Varanasi and Sarangi (1975). They reported an N value of 0.75, in agreement with the theoretically predicted value, and they noted the need for further studies. [Work above room temperature by Sell (1980) has been commented on by Varanasi (1981).]

Equation (38) is very important to several methods of sensing temperature by FT-IR spectrometry where the application of Eqs. (16) and (24) requires the calculation of good values of γ_c and hence an accurate knowledge of N. More recently the temperature-broadening exponent N has been measured as 0.69 by Lowry and Fisher (1982) for the P(3), P(7), and P(12) lines of CO broadened by N_2, and an average value of 0.66 has been measured by Sun and Griffiths (1981) for the P(26), P(27), P(29), and P(31) lines of $^{12}C^{16}O$, and the P(9), P(13), and P(16) lines of $^{13}C^{16}O$. Bouanich et al. (1983) measured N for N_2-broadened lines of CO at low temperatures, and although they found no significant dependence of N on m, N was found to vary with temperature from 0.73 at 190 K to 0.765 at 93 K. They stated that a value of 0.66 for N at room temperature was calculated from their low-temperature data. The measurement of Griffiths and Sun, the method of which is reviewed in this chapter, was on self-broadened CO, while the other values of N given above are for

CO broadened by N_2 ($CO-N_2$). Recently Nakazawa and Tanaka (1982b) have reported values of *N* for a set of self-broadened lines of the CO fundamental mode, R(1) through R(17). They found that *N* reaches a maximum of 0.778 at R(7) and then decreases monotonically with increasing rotational index *m* to a value of 0.542 for the R(17) line.

The variation in *N* which has been observed is predicted by Ehrenfest's adiabatic principle, which applies to the energy coupling which may occur between some perturbation, in this case the colliding molecule, and some quantized observable which it acts upon, such as the rotational energy (Lambert, 1977). At high values of the rotational quantum number *J*, the period of a single rotation becomes small in comparison to the duration of a collision, so energy transfer between the colliding and the active molecule is lessened, essentially making the collision more elastic. The value of *N* expected for elastic collisions between hard spheres is 0.5, as noted earlier in this section. Estimating temperature from moderate resolution spectra requires values for γ_c for each of the lines; so the dependence of *N* on *m* is very important. Nakazawa and Tanaka also found different values for *N* when either O_2 or N_2 was used as the broadening gas. To apply the spectroscopic temperature probe to samples in an air atmosphere, the ($CO-N_2$) and ($CO-O_2$) values of *N* should be used to predict γ_c. For combustion gases, ($CO-CO_2$) and ($CO-H_2O$) data are also required (Varghese and Hanson, 1981a). Further practical measurements of the temperature dependence of collision-broadened linewidths appear to be vitally important.

5.4 Voigt Profile

At pressures between a few Torr, below which lines are Doppler-broadened, and 100 Torr, where lines are almost entirely collision-broadened, the line profile is determined by a convolution of the Gaussian and Lorentzian shapes. The convolution of two functions, $f(\bar{\nu})$ and $g(\bar{\nu})$, each having finite integrals, is given by

$$h(\bar{\nu}_0) = \int_0^\infty f(\bar{\nu})g(|\bar{\nu} - \bar{\nu}_0|)\, d\bar{\nu} \qquad (39)$$

The convolution of Gaussian and Lorentzian functions is known as a Voigt function. Several approaches have been taken to simulate the Voigt function, which may be expressed as

$$K(\xi,a) = \frac{\kappa_{\bar{\nu}}}{K'} = (a/\pi)\int_0^\infty \frac{\exp(-y^2)}{a^2 + (\xi - y)^2}\, dy \tag{40}$$

where K' is the dimensional constant of the reduced absorption coefficient, a is proportional to the ratio of γ_c to γ_D, and ξ is the wavenumber scale in units of Doppler width; they are given explicitly by

$$K' = \frac{S}{\gamma_D}\left(\frac{\ln 2}{\pi}\right)^{1/2} \tag{41}$$

$$a = \frac{\gamma_c}{\gamma_D}(\ln 2)^{1/2} \tag{42}$$

$$\xi = \frac{\bar{\nu} - \bar{\nu}_0}{\gamma_D}(\ln 2)^{1/2} \tag{43}$$

The variable y is a dummy variable, because it integrates out of the Voigt profile, which is a function of the reduced frequency ξ. The infinitesimal dy is the width of a narrow rectangular slice of height $[a^2 + (\xi - y)^2]^{-1}$ at each value of y. Each of these narrow rectangles, at each value of ξ, is broadened by the Gaussian factor $\exp(-y^2)$. The integral sums up the contribution from each rectangle, so the Voigt profile is a convolution of the Lorentzian factor, $(a^2 + (\xi - y)^2)^{-1}$ with the Gaussian function. Intuitively the Lorentzian profile can be considered to be "smeared out" by the error function, $\exp(-y^2)$. Various methods of computing Voigt functions have been published (Noll and Pires, 1980; Drayson, 1976; Kielkopf, 1973; Klim, 1981).

To calculate the Voigt parameter a for different temperatures using Eq. (42), the temperature dependence of both γ_D and γ_c must be known. γ_D is known to vary as $T^{1/2}$, but the temperature dependence of γ_c still requires study in individual cases. In Sec. 6, the results of a detailed study of the temperature dependence of γ_c will be described for a particular case.

6 TEMPERATURE DEPENDENCE OF γ_c

6.1 Self-Broadening

The necessity of an accurate knowledge of the temperature dependence of collision-broadened linewidths, or pressure broadening coefficient, led us to initiate a series of measurements to study this parameter. A TDL spectrometer was modified to permit the rapid acquisition of digital data (Sun et al., 1980), and high resolution spectra of carbon monoxide were measured both self-broadened and with added nitrogen. The temperature of the gas was varied between 297 and 453 K.

For the determination of N, Eq. (38) was rearranged to give the pressure broadening coefficient γ_0:

$$\gamma_0 = \frac{\gamma_c}{p} = \Gamma T^{-N} \tag{44}$$

where

$$\Gamma = \left(\frac{\gamma^0}{p_0}\right) T_0^N \tag{45}$$

Taking logarithms:

$$\ln \gamma_0 = \ln \Gamma - N \ln T \tag{46}$$

The pressure broadening coefficient γ_0 can be measured over a series of temperatures by measuring the half-width at half-height of a rotational line γ_H at different pressures and correcting for Doppler broadening through Whiting's approximation (Whiting, 1968):

$$\gamma_H = \frac{\gamma_c}{2} + \left[\frac{\gamma_c^2}{4} + \gamma_D\right]^{1/2} \tag{47}$$

γ_H was found by fitting the measured spectral line to a Lorentzian function using a nonlinear least squares regression technique. The pressures being used in this work were always sufficiently high that even if the lines were not exactly Lorentzian, the Voigt profile was more similar to a Lorentzian than a Gaussian function. For CO above 200 Torr, the Doppler contribution is negligible.

Typical spectra observed after increases in pressure are shown in Fig. 4. It can be seen that at low pressures, where the contribution of Doppler broadening is still significant, both A_{peak} and γ_H increase as the pressure is increased. Above about 200 Torr, the peak absorbance remains constant with in-

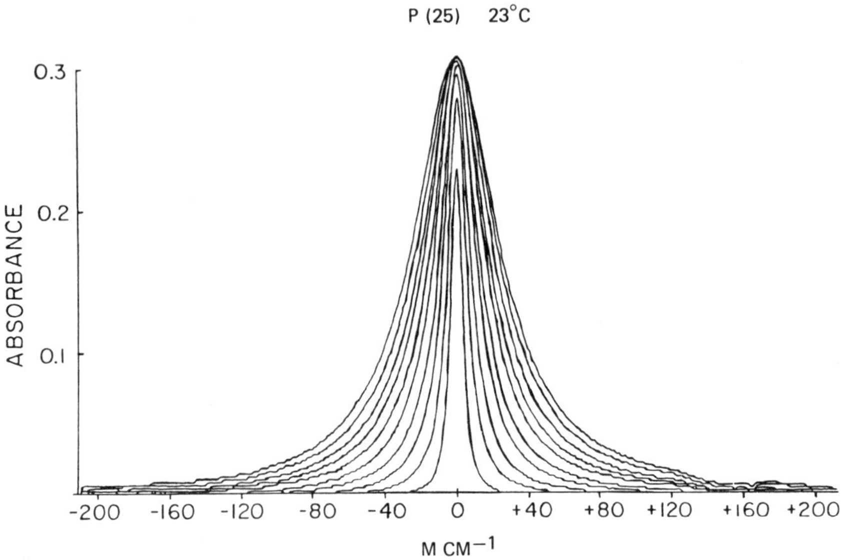

Fig. 4 Absorbance spectra of the P(25) line of CO measured using a TDL spectrometer at pressures from 50 to 500 Torr, at 50-Torr intervals (from Sun and Griffiths, 1981).

creasing pressure, and the half-width increases linearly with pressure. In this regime, collision broadening is the dominant mechanism and the profile may be quite accurately represented by a Lorentzian function. A typical plot of γ_c against pressure for temperatures between ambient and 453 K is shown in Fig. 5, where γ_c has been calculated using Eq. (47). Self-broadening coefficients could be determined from plots of this type with a relative standard deviation of approximately 1%.

The effect of temperature on the linewidth and peak absorbance at constant pressure is illustrated in Fig. 6. A_{peak} increases dramatically as the Boltzmann population of the higher rotational energy states increases as the temperature is raised. When the values of $\ln \gamma_0$ are plotted against $\ln T$, good linearity is usually observed. One of the better plots is shown in Fig. 7, for which the pressure broadening coefficients for the P(27) line of CO calculated from the data in Fig. 5 were used.

The average value for N found from the P(26), P(27), P(29), and P(31) lines of $^{12}C^{16}O$ and the P(9), P(13), and P(16) lines of

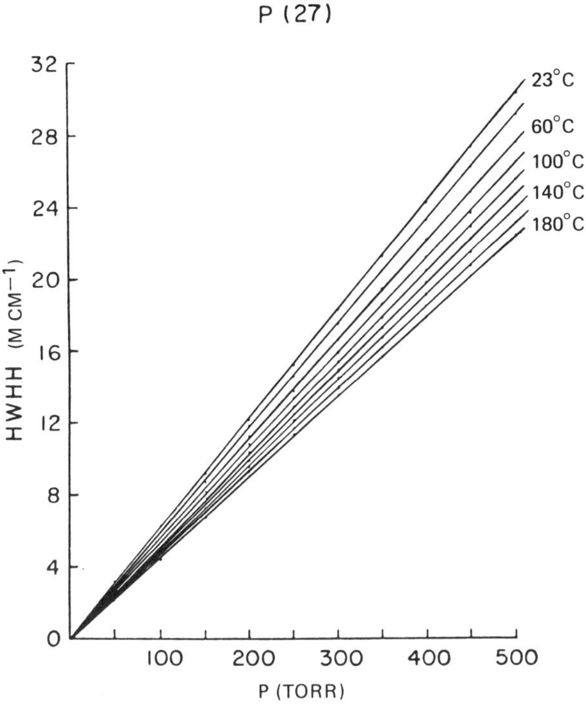

Fig. 5 Series of plots of γ_C against pressure for the P(27) line of self-broadened CO at temperatures from 296 K (upper line) to 453 K (lower line) at intervals of 20 K (from Sun and Griffiths, 1981).

$^{13}C^{16}O$ was 0.66, with the standard deviation due to measurement errors being 0.04 (Sun and Griffiths, 1981). Although these data represented the most accurate values for N which had been reported at that time for the fundamental band of CO, they were not sufficiently accurate to indicate whether or not the value of N varied with m, the index of the line. Nevertheless the values of N reported for each of the seven lines listed above was significantly less than the value of 0.75, which had previously been generally accepted.

There are several reasons why only a few lines were studied in this work. First, the lines of low J'' have very large line strengths. The peak absorbance of the strongest of these lines

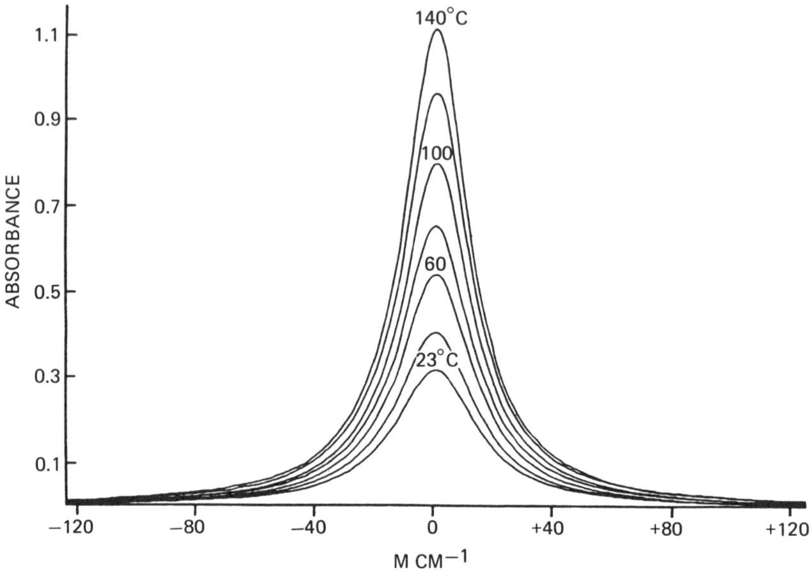

Fig. 6 The effect of temperature on the linewidth and peak absorbance of the P(25) line of CO at 300 Torr pressure. The absorbance increases because the $J'' = 25$ level becomes more populated with increasing temperature.

measured at 760 Torr over a path length of 10 cm may be over 450, far exceeding the limits allowed by the SNR of the spectrum or the dynamic range of the data acquisition system. Second, the laser diode being used did not permit any line in the R branch to be studied, and even for the P branch it was often found that no combination of diode temperature and current permitted a mode to be isolated for certain lines. A single mode is necessary for the output wavenumber to be proportional to the diode current. Finally, severe interferences were occasionally caused by the presence of overlapping lines due to other isotopic species (e.g., $^{13}C^{16}O$ or $^{12}C^{18}O$) or from hot bands when measurements were made with the sample at a high enough temperature that the $\nu = 1$ vibrational state becomes populated significantly.

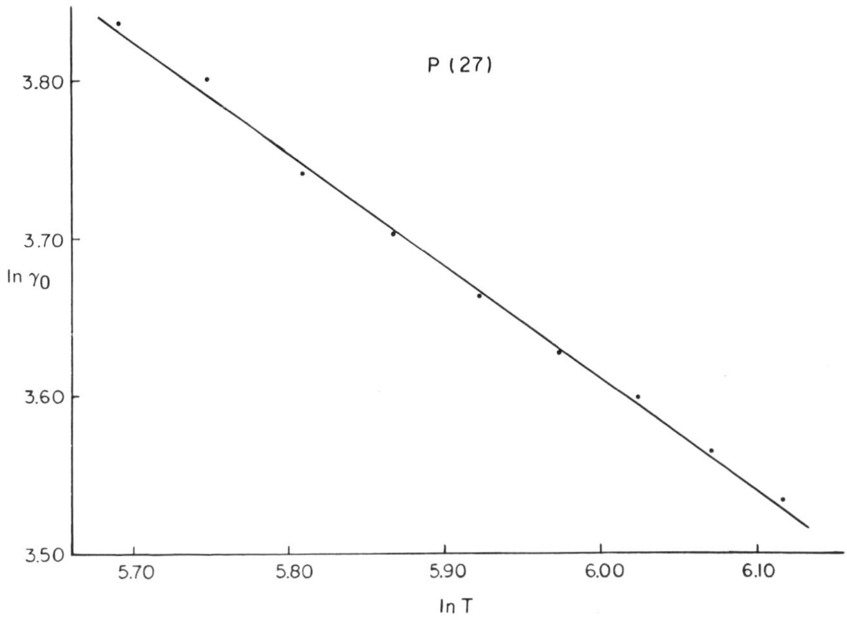

Fig. 7 Plot of ln γ_0 against ln T for the P(27) line of CO. The value of N given by the slope of this plot is 0.61 (from Sun and Griffiths, 1981a).

6.2 Foreign-Gas Broadening

For many years it has been believed that the effects of self-broadening and foreign-gas broadening are additive (Darnton and Margolis, 1973). If this were the case, Eq. (29) would be valid. For carbon monoxide in the presence of nitrogen, the collision-broadened linewidth at a standard temperature T_0 would be given by

$$\gamma = \gamma^0_{CO}\left(\frac{P_{CO}}{760}\right) + \gamma^0_{CO,N_2}\left(\frac{P_{N_2}}{760}\right) \tag{48}$$

where P_{CO} is the partial pressure of CO (Torr), P_{N_2} the partial pressure of nitrogen, γ^0_{CO} the self-broadening coefficient of CO, and γ^0_{CO,N_2} the nitrogen-broadening coefficient of CO.

The effect of measuring at a temperature T, instead of the standard temperature T_0, could be included as

$$\gamma = \gamma^0_{CO}\left(\frac{P_{CO}}{760}\right)\left(\frac{T_0}{T}\right)^{N_{CO}} + \gamma^0_{CO,N_2}\left(\frac{P_{N_2}}{760}\right)\left(\frac{T_0}{T}\right)^{N_{N_2}} \qquad (49)$$

A series of measurements of γ_c with differing amounts of CO in N_2 were made to test for agreement with Eq. (48), and the results were surprising (Sun and Griffiths, 1981a). Shown in Fig. 8 are the plots of γ_c versus pressure for the P(25) line of CO alone, in mixtures of 1% CO in N_2, and 50% CO in N_2. For the mixture of 1% CO in N_2 the broadening is almost entirely due to collisions with N_2 molecules and therefore the slope of this plot is the value of γ^0_{CO,N_2} in Eq. (49). Similarly, the slope for the self-broadened CO plot is the value of γ^0_{CO}. A mixture of 50% CO has partial pressures of CO and N_2, which are each equal to one half of the total pressure. Substitution into Eq. (49) predicts that the slope of the best fit for the 50% mixture should be the average of γ^0_{CO} and γ^0_{CO,N_2}. Figure 8 shows that the actual slope is less, since the plot for the 50% mixture is closer to that of the 1% mixture than to the plot for self-broadened CO.

No other observations on this effect are known, and it is not known whether a similar result would be observed if a different broadening gas were used. It is of interest to study if the effect occurs in general or if it is a result of a particular set of experimental circumstances such as the choice of the line, the path length of the cell, or the sample temperature. In Fig. 8, for example, at 413 K the plot of γ_c for the 50% mixture lies nearly midway between those of the self- and N_2-broadened plots.

6.3 Comparison of Results for Linewidth

Low resolution measurements employing curve-of-growth methods for linewidth determinations have been made for both branches of the 1–0 band of CO (Draegert and Williams, 1968; Bouanich and Haeusler, 1972). The variation of linewidth with line index m has been observed in these studies and verified by direct, high resolution measurements (Hanson, 1983).

Further high resolution studies of the CO fundamental band have been made (Lowry and Fisher, 1982), and the results are in good agreement with those of Varghese and Hanson (1980). A comparison of the data from several studies is shown in Fig. 9, in

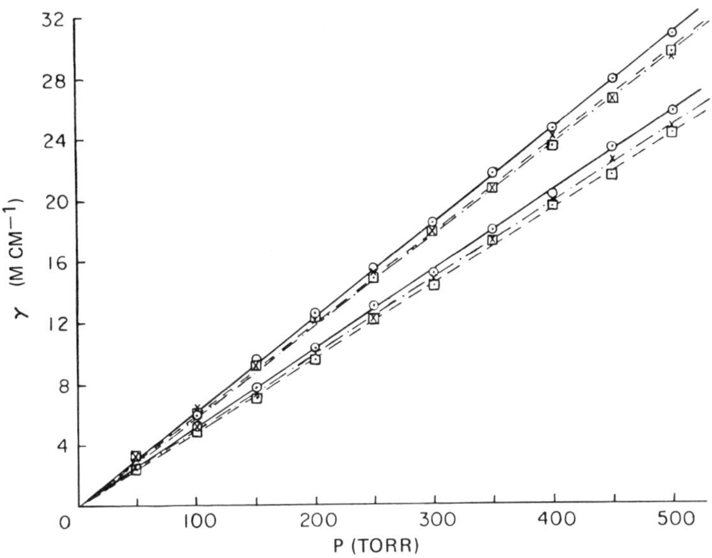

Fig. 8 Plots of the collision-broadened half-width of the P(25) line of CO against total pressure for 100% CO (⊙), 50% CO in N_2 (×), and 1% CO in N_2 (⊡) at temperatures of 313 K (upper set) and 413 K (lower set).

which the reported values of γ_c^0 (the half-width at standard temperature and pressure) are plotted as a function of line index *m*. The results of measurements on self-broadened CO and CO broadened by N_2, CO_2, and H_2O are listed in Table 2.

The data in Table 2 have been standardized to a temperature of 300 K and pressure of 1 atm. Lowry et al. (1982) note that in contrast to the results for CO–N_2 and CO–CO_2, the values reported for CO–H_2O were not in very good agreement with the ones previously reported for this mixture (Varghese and Hanson, 1981a). A possible explanation for this observation which was given by Lowry et al. was that their value was extrapolated from higher temperature data, while the previous result was obtained at room temperature and a lower pressure (25 Torr).

The direct measurement of the temperature broadening exponent is also discussed in this report. The same lines were measured over a temperature interval from ambient to 600 K, and

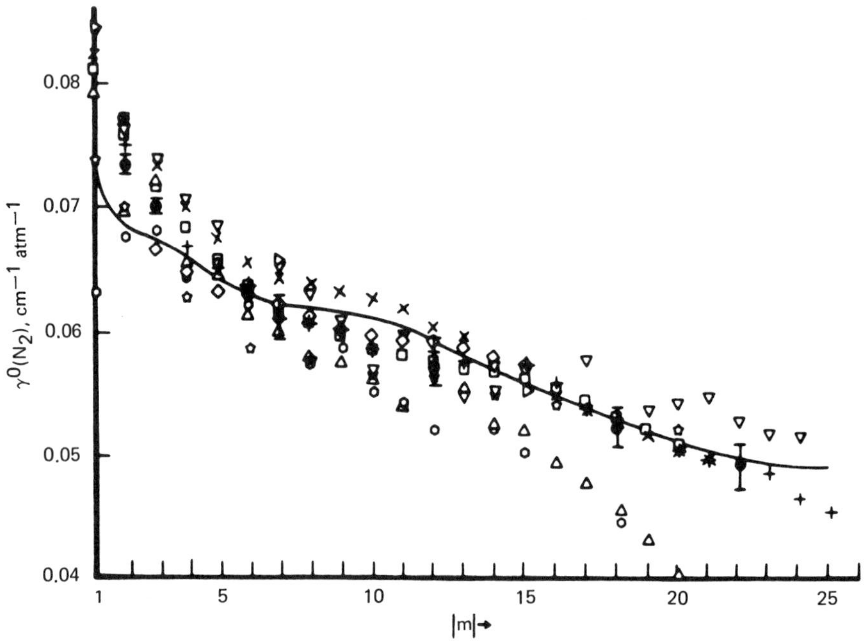

Fig. 9 Linewidths of the N_2-broadened CO fundamental band plotted as a function of line index *m*. The solid line represents the values calculated using the Anderson–Tsao–Curnutte theory, and the symbols are identified as follows: ●, Lowry and Fisher (1982); ♦, Varanasi and Sarangi (1975); +, Bouanich and Farrenq (1970); △, *R*-branch and ⬡, *P*-branch, Crane-Robinson and Thompson (1963); X, Bouanich and Brodbeck (1973); □, Bouanich and Haeusler (1972); ◊, *ibid.* for (3,0) band; ○, Draegert and Williams (1968); ⬠, James and Plyler (1964); ▽, Varghese and Hanson (1980) (from Lowry and Fisher, 1982).

one line, P(7), was measured from 300 to 1000 K. Their results for the lower J lines agree to within experimental error with the values reported by Sun and Griffiths (1981) for the P(26) and P(27) lines, and the values given in their report are listed in Table 3.

Comparing measurements of *N* made in different laboratories is difficult, because the lines which have been selected, the broadening gases and their proportions, and the temperature ranges

TABLE 2 CO Line Strengths and Linewidths at 1 Atm Measured by Lowry et al. (1982)

Line	Line strength S^{exp}, $\frac{cm^{-2}}{atm}$	γ^0 (300 K) $\times 10^3$ cm^{-1}			
		CO–CO[a]	CO–N_2	CO–CO_2	CO–H_2O
P(2)	4.388	83.1	74.77	99.99	99.16
P(3)	5.857	78.6	69.55	87.55	92.14
P(7)	9.410	66.0	60.69	73.40	99.56
P(12)	6.384	64.5	56.94	67.14	
P(18)	1.728	57.2	52.72		
P(22)	0.467		49.91		

[a]Self-broadening data obtained by Nakazawa and Tanaka (1982a).

chosen, all differ. It is likely that N may itself be found to vary with temperature if results measured at very high or very low temperature are compared. In addition, several other laboratories have also reported that the exponent N varies with the index of lines m. Varghese and Hanson (1981b) have plotted this variation based on their results of studies of combustion gases measured using a TDL spectrometer. The trend to slightly smaller values of N with increasing index m has also been observed by Nakazawa and Tanaka (1982b), using a grating spectrophotometer through the application of a curve-of-growth technique.

7 EXPERIMENTAL MEASUREMENTS OF TEMPERATURE

7.1 Band Fitting

In order to test the application of the ROTMP programs described earlier in this chapter, a series of absorption and emission spectra of carbon monoxide, both self-broadened and nitrogen-broadened, were measured using a Fourier transform spectrometer operating at 0.2 and 0.5 cm^{-1} nominal resolution with triangular apodization. The actual value of resolution is necessary for the calculation of the resolution parameter ρ. When CO was held at less than 10-Torr pressure, the lines are Doppler-broadened

TABLE 3 The Exponent *N* as Measured by Lowry et al. (1982) for N_2-Broadened CO

Line	*N*
P(2)	0.62
P(3)	0.69
P(7)	0.69
P(12)	0.69

(with $\gamma_D \cong 0.002\ cm^{-1}$), so that the measured line profile should be approximately the same as the ILS function. The FWHH of the lines in the spectrum of CO at low pressure measured at 0.2-cm^{-1} resolution was 0.25 ± 0.03 cm^{-1}, which is the FWHH of an FT-IR spectrometer operating at 0.2-cm^{-1} nominal resolution and triangular apodization. The known value of δ_0 could then be entered into Eq. (21) for the calculation of ρ. A 1.0-cm path-length cell was used for measurements on self-broadened CO because of the intense line strength of carbon monoxide. In the studies of N_2-broadened carbon monoxide, CO at a concentration of 1% in N_2 was held in a 10-cm gas cell. The pressure of the gas in the cell could be controlled to within ±1 Torr. Absorption spectra were measured at temperatures from ambient to 180°C (453 K), with the temperature of the cell being controlled to ±1°C. An example of a single-beam transmittance spectrum of the CO band at 0.2-cm^{-1} resolution is shown in Fig. 10. Interfering lines primarily due to ^{13}CO can be observed in the P branch of the band.

Because the energy emitted by a blackbody is very small at temperatures close to ambient, only temperatures above 373 K were used in the emission study. In this case, the emission spectrum of the empty cell was always subtracted from the emission spectrum of the cell with the sample present. To ensure the highest possible photometric accuracy, all interferograms were zero-filled (Griffiths, 1975b) by a factor of 4.

Absorption Spectrometry

Very accurate estimates of the temperature of CO could be obtained when the gas is at ambient temperature. For spectra measured at either 0.2- or 0.5-cm^{-1} resolution, the calculated temperature was usually within 2 K of the measured value, and often

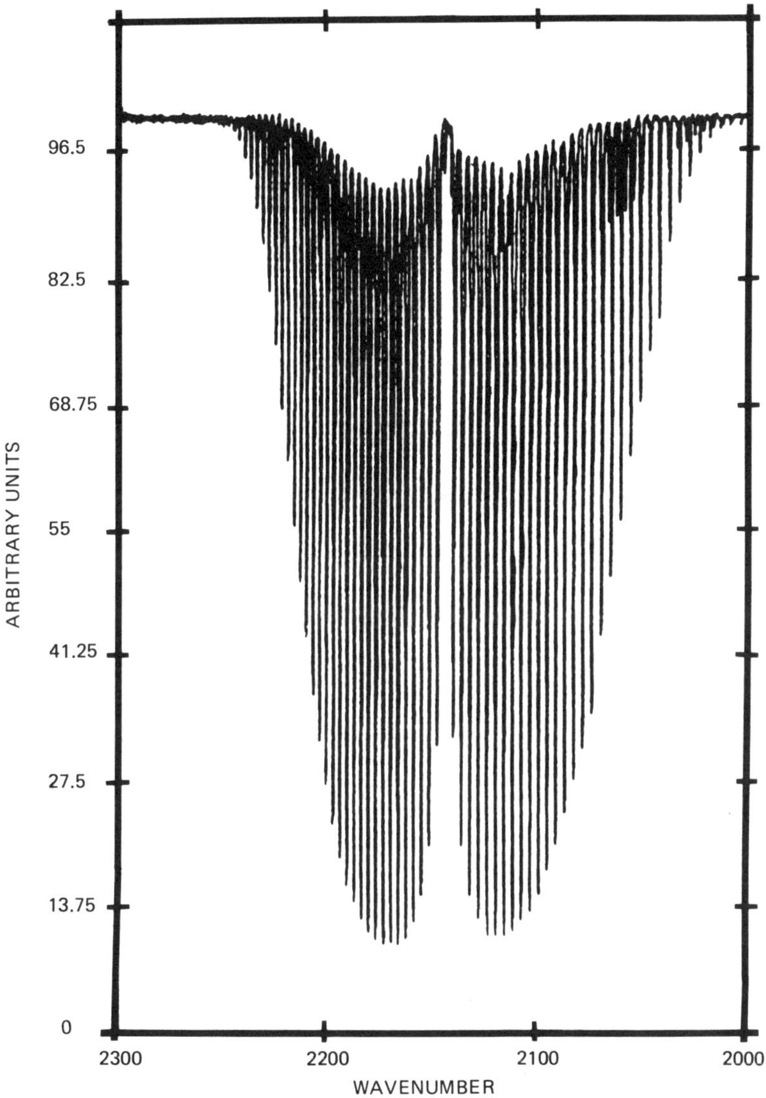

Fig. 10 Transmittance spectrum of the fundamental band of CO at 500 Torr and 298 K measured at 0.2-cm^{-1} resolution. The cell pathlength is 1 cm.

within 1 K. These results indicate that the ROTMP program yields accurate temperatures when all the input parameters are correct. Compatible selected values from the linewidth data of Bouanich and Haeusler (1972), Bouanich and Brodbeck (1973), and Tejwani (1972) were input to the ROTMP program, since they appeared to correspond best to our measured values for lines of high index measured using a TDL spectrometer.

As the temperature of the gas was increased, the temperature calculated using the ROTMP program deviated increasingly from the measured temperature, until at 453 K the error was about 20°. Equal errors were found for spectra measured at 0.2- or 0.5-cm^{-1} resolution. Thus the effect of spectrometer resolution appears to be accurately compensated in the ROTMP program. Lack of accurate knowledge of the temperature dependence of the self-broadening coefficient was the most likely factor leading to the errors in the temperature estimation. Changing the value of N used in the ROTMP program between 0.66 and 0.75 only led to a very small change in the calculated temperature. Only by assuming that N varies with the line index m could slightly better data be obtained.

Several different empirical equations for the dependence of N on m have been tried, and the estimated temperature retrieved using each of these functions has been compared to the actual temperature, 453 K at 500 Torr of pure CO, in Table 4. It can be seen that the estimated temperature is quite sensitive to the way in which N is assumed to vary with m. It has also been shown that by using $N = -0.001\,m + 0.75$, the estimated temperature exactly matches the experimental temperature. Recalling that the value of N for elastic collisions between hard spheres is 0.5, it is reasonable to expect N to decrease as the rate of molecular rotation increases. At high values of $|m|$ (by definition, high values of the rotational quantum number J) the anisotropic forces which arise during collisions are averaged out, in effect making collisions more elastic. For spectra measured with CO at other temperatures and pressures, the results using $N = -0.001\,m + 0.75$ are shown in Table 5. The agreement between the actual and estimated temperatures in this table indicates that the variation of N with m is the factor limiting the accuracy of the overall method.

The approximate formula given for N was subsequently applied to FT-IR spectra of CO and N_2 mixtures in the 10-cm gas cell. At each temperature spectra were obtained for samples containing 1% CO in N_2 at total pressures of 500 and 760 Torr. Referring to Table 6, where estimated temperatures are shown for only the 0.2-cm^{-1} spectra, it is clear that the accuracy is poor when the

TABLE 4 Results of Study into the Dependence of N on $|m|$ for Self-Broadened CO at 453.2 K and 500-Torr CO from the Spectrum Measured at 0.2-cm^{-1} Resolution

Equation	T	A_1
$N = 0.75$	474.1	0.920
$N = 0.0002\ m + 0.75$	469.3	0.902
$N = 0.0004\ m + 0.75$	472.2	0.917
$N = 0.001\ m + 0.75$	480.8	0.964
$N = -0.004\ m + 0.75$	461.1	0.860
$N = -0.001\ m + 0.75$	453.1	0.820
$N = -0.0005\ m + 0.75$	459.8	0.853
$N = -0.003\ m + 0.75$	428.5	0.698
$N = 0.66$	475.9	0.859
$N = 0.0004\ m + 0.66$	494.7	0.398
$N = 0.001\ m + 0.66$	495.2	0.458
$N = -0.005\ m + 0.66$	518.2	0.283
$N = 0.000026\ m^2 - 0.0001\ m + 0.75$	464.8	0.849
R: $N = -0.002\ m + 0.75$		
P: $N = -0.003\ m + 0.75$	446.6	0.818

sample is foreign-gas-broadened. As was the case with the pure CO samples, the error between the estimated and actual temperature is small at ambient temperature, and with increasing temperature the error becomes progressively greater. Because the results follow a similar pattern, it is felt that an improved knowledge of the variation of the exponent N as a function of m will permit more accurate rotational temperatures to be estimated for nitrogen-broadened CO. Experiments to establish this function empirically are in progress.

Emission Measurements

An example of an emission spectrum of nitrogen-broadened CO is shown in Fig. 2. Certain corrections must be made to the spec-

TABLE 5 Estimated Temperatures for Self-Broadened CO with Different Dependence of N on $|m|$ for P and R Branches

			$N = -0.001\,m + 0.75$		R: $N = -0.001\,m + 0.75$ P: $N = -0.003m + 0.75$	
Sample	T(expt) (K)	P (Torr)	0.2 cm^{-1} T (calc) (K)	0.5 cm^{-1} T (calc) (K)	0.2 cm^{-1} T (calc) (K)	0.5 cm^{-1} T (calc) (K)
1	299	300	299.3	293.8	297.6	290.0
2	299	500	293.3	292.1	291.9	288.7
3	333	300	338.3	327.8	336.4	323.0
4	333	500	327.6	325.7	326.2	321.3
5	373	300	378.3	369.2	375.3	362.8
6	373	500	371.1	365.2	368.1	359.2
7	393	300	401.5	394.2	397.7	388.3
8	393	500	389.4	386.7	385.4	379.6
9	413	300	422.6	415.4	417.9	406.9
10	413	500	411.3	409.1	406.2	401.0
11	433	300	447.4	437.9	441.8	428.5
12	433	500	435.5	431.4	429.5	422.4
13	453	300	467.8	464.5	461.5	453.9
14	453	500	453.1	451.7	446.6	441.9

trum as measured to obtain comparable intensities to a transmittance spectrum such as that shown in Fig. 10. The measured single-beam spectrum is essentially the sum of the energy reaching the detector from several sources, including the gas sample, the cell, and the interferometer optics. The signal from the latter two sources was removed from the spectrum by subtracting the spectrum of the heated, evacuated cell from the measured spectrum. The efficiency of the interferometer optics and the response of the detector also vary over the wavenumber interval of the band. To correct for these effects, a spectrum was obtained of a laboratory hot plate painted black which was placed

TABLE 6 Estimated Temperatures for Nitrogen-Broadened CO with Different Dependence of N on $|m|$ for the P and R Branches

			$N = -0.001m + 0.75$		R: $N = -0.001m + 0.75$ P: $N = -0.003m + 0.75$	
Sample	T (expt) (K)	P (Torr)	0.2 cm^{-1} T (calc) (K)	0.5 cm^{-1} T (calc) (K)	0.2 cm^{-1} T (calc) (K)	0.5 cm^{-1} T (calc) (K)
1	298	500	301.5	299.6	295.0	292.7
2	298	760	301.1	298.6	296.1	292.7
3	333	500	336.8	331.3	329.2	323.0
4	333	760	339.0	332.3	332.7	325.0
5	373	500	384.1	377.5	374.6	367.3
6	373	760	380.2	377.2	372.4	368.1
7	393	500	429.5	399.5	418.3	388.6
8	393	760	339.3	399.8	390.5	389.6
9	413	500	451.6	421.6	439.2	409.1
10	413	760	424.6	420.9	414.9	409.6
11	433	500	478.2	443.5	464.1	430.1
12	433	760	452.7	441.4	441.3	428.8
13	453	500	498.9	463.0	483.5	449.9
14	453	760	466.1	463.6	453.6	449.4

behind two KBr windows to ensure temperature uniformity. The emission spectrum of this source approximates that of a blackbody. Using this spectrum, the instrument response function (IRF) could be measured. The IRF is the ratio of the measured blackbody spectrum to the theoretical spectrum calculated over the interval of the band. Ratioing the background-subtracted spectrum of the hot gas to the instrument response function results in the corrected spectrum shown in Fig. 2. The peak intensities for this spectrum provided the input data for the ROTMPE program.

The results of the emission studies are summarized in Tables 7 and 8. It was found that the emission results give approximately the same magnitude of error as was found in the absorption studies. The temperature difference at 473 K is again about 20 K for CO self-emission, suggesting that no further errors are introduced in the ROTMPE program used for emission spectrometry. As for the corresponding results for absorption spectrometry, it would be expected that this difference would be reduced if the variation of *N* with the index of the line were known more accurately. For N_2-broadened CO emission the estimated temperature is close to the experimental temperature except for the emission at 373 K. At 373 K, the emitted energy for CO at low pressure is too low to permit a good signal-to-noise ratio to be obtained. Extensive signal averaging would have to be done in order to improve the SNR. For a sample of 373 K measured at 0.2-cm^{-1} resolution, the SNR for fairly strong lines is about 500. It should be noted that a data acquisition time of 30 min was required to achieve this sensitivity, but the SNR might be improved by using lower resolution or more extensive signal averaging. The capability of FT-IR spectrometry to achieve high SNR and the necessary accuracy in the measured peak intensities through signal averaging is a great advantage for the purpose of probing temperature spectroscopically, unless very fast data acquisition times are required.

7.2 Graphical Methods

Few results of temperature determinations using the graphical methods described earlier have been reported. Anderson and Griffiths (1977) reported temperatures calculated in several different ways from spectra of a laboratory sample of carbon monoxide measured using an FT-IR spectrometer. For CO at 108 Torr held in a 10-cm gas cell at ambient temperature, the peak absorbance of the strongest lines, $-\ln \tau(m)$, was much greater than 25. Since the spectrum was measured at relatively low resolution ($1\ cm^{-1}$), the square root dependence shown in Eq. (25) holds for all lines for which $E(m)$ is less than $1400\ cm^{-1}$.

Shown in Fig. 11 are plots of three functions of $A^a_{peak}(m)$ plotted against $E(m)$. In the lowest plot, $\ln[(A^a_{peak})/|m|F(m)\bar{\nu}(m)]$ is plotted against $E(m)$, i.e., the effects of the variation of linewidth with *m* and resolution errors are neglected. A distinct curvature is seen in this plot, and the calculated temperature is in error by +48%. In general when the effect of finite resolution is neglected, the points on the graph at low $E(m)$ often fall off

TABLE 7 Estimated Temperatures Calculated from Emission Spectra of Self-Broadened CO, with $N = 0.75$ and $N = 0.66$

			$N_1 = 0.75$		$N_1 = 0.66$	
Sample	(K)	P (Torr)	0.2 cm^{-1} T (calc) (K)	0.5 cm^{-1} T (calc) (K)	0.2 cm^{-1} T (calc) (K)	0.5 cm^{-1} T (calc) (K)
1	453	300	480.7	498.6	480.8	497.8
2	453	500	478.3	467.6	479.8	473.2
3	473	300	496.9	498.9	498.0	498.3
4	473	500	493.4	497.1	496.8	488.5

the straight line. For the upper plot, the ordinate is $\ln[(A^a_{peak}(m))^2/|m|F(m)\bar{\nu}(m)]$, i.e., only the effect of variation of γ with m is neglected. In this case, the temperature calculated from the slope (for $E(m) < 1400$ cm^{-1}) was in error by -7.3%. In the center curve, the ordinate is $\ln[(A^a_{peak})^2/|m|F(m)\gamma(m)\bar{\nu}(m)]$ and the calculated temperature was in error by only 0.4%. The need for an accurate knowledge of the variation in γ with respect to m is apparent from these results. This very good result was obtained mainly because the values of $\gamma(m)$ are known to high accuracy at ambient temperature. At higher temperatures the error increases, with errors of about -5% being found for gases at about 400 K when ambient temperature values for $\gamma(m)$ were used. Presumably this error was again due to an inaccurate knowledge of the temperature exponent N.

Practical observations of the emission spectra of carbon dioxide from industrial stack plumes have been reported by Herget and Brasher (1980). Their spectra are reproduced in Fig. 12, both for a carbon monoxide boiler and a coal-burning power plant measured at distances of 600 m and 800 m, respectively. Spectra were measured at 0.12 cm^{-1}, so that the $\boldsymbol{\rho}$ values were approximately equal to unity for each line. When $\ln[\bar{\nu}(m)^4/\gamma(m)I(m)]$ was plotted against $E(m)$, linear plots were obtained in both cases, as illustrated in Figs. 13 and 14. The calculated temperatures were in remarkably good agreement with the temperatures measured by in-stack monitors.

Perhaps a few caveats are in order concerning these results. First of all, it is likely that the emissivities of many of the lines

TABLE 8 Estimated Temperatures Calculated from Emission Spectra of Nitrogen-Broadened CO, with $N = 0.75$ and $N = 0.66$

			$N = 0.75$		$N = 0.66$	
Sample	T (expt) (K)	P (Torr)	0.2 cm^{-1} T (calc) (K)	0.5 cm^{-1} T (calc) (K)	0.2 cm^{-1} T (calc) (K)	0.5 cm^{-1} T (calc) (K)
1	373	500	283.6		284.2	
2	373	760	360.5		361.6	
3	423	500	417.2	423.5	419.6	424.8
4	423	760	435.1	419.3	436.6	421.6
5	473	500	480.4	489.8	482.7	490.6
6	473	760	492.3	488.0	493.4	489.3

are sufficiently large that $I(m)$, which is proportional to $[1 - \tau(m)]$, is no longer even approximately proportional to the absorbance. An example of this was seen in Chap. 1 in comparing Figs. 17 and 18. Since $\rho \cong 1$, there should be a finite resolution error, especially for the stronger lines in the spectrum, as indicated by the deviation of the first point in Fig. 13. In addition, the P and R branches should yield identical temperatures. Since it is likely that the plume temperature would be somewhat less than the in-stack temperature, it is quite possible that the agreement between the calculated and in-stack temperatures is fortuitous. Nevertheless these data definitely indicate the feasibility both of measuring high quality remote emission spectra at quite high resolution with a Fourier spectrometer, and of deriving at least fairly accurate temperatures from these spectra. Much more work using this method is apparently called for. However it should be recognized that obtaining a linear plot is usually the result of one of several criteria being fulfilled. In general we believe that band-fitting methods should yield more accurate temperatures, since fewer assumptions or approximations are required.

7.3 Two-Line-Method Temperature Measurements

The two-line method differs from the temperature sensing methods which use the entire band. Since two adjacent lines must be mea-

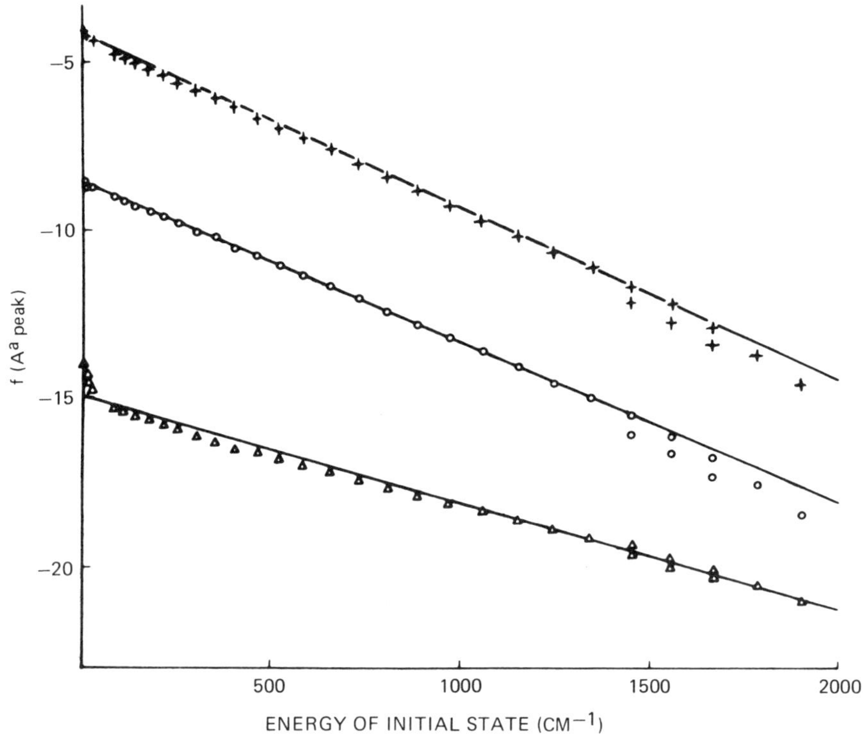

Fig. 11 Comparison of temperature analysis plots for CO spectra measured at moderate resolution using three different functions of the apparent absorbance, $f(A^a_{peak})$. These three graphical methods use different approximations (see text for details). For clarity, the ordinates of the three plots have been offset by adding or subtracting a constant factor of 2 (from Anderson and Griffiths, 1977).

sured, a high resolution TDL spectrometer may be used to record the spectral data. The present state of development of these instruments requires that the measurement of temperature be made by an absorption technique, although lasers are being developed with higher power, which will make emission spectroscopy possible through heterodyne methods (Glenar, 1983). The two-line method for measuring temperature has been used for many years (Penner, 1959), although the application of this method to lines in

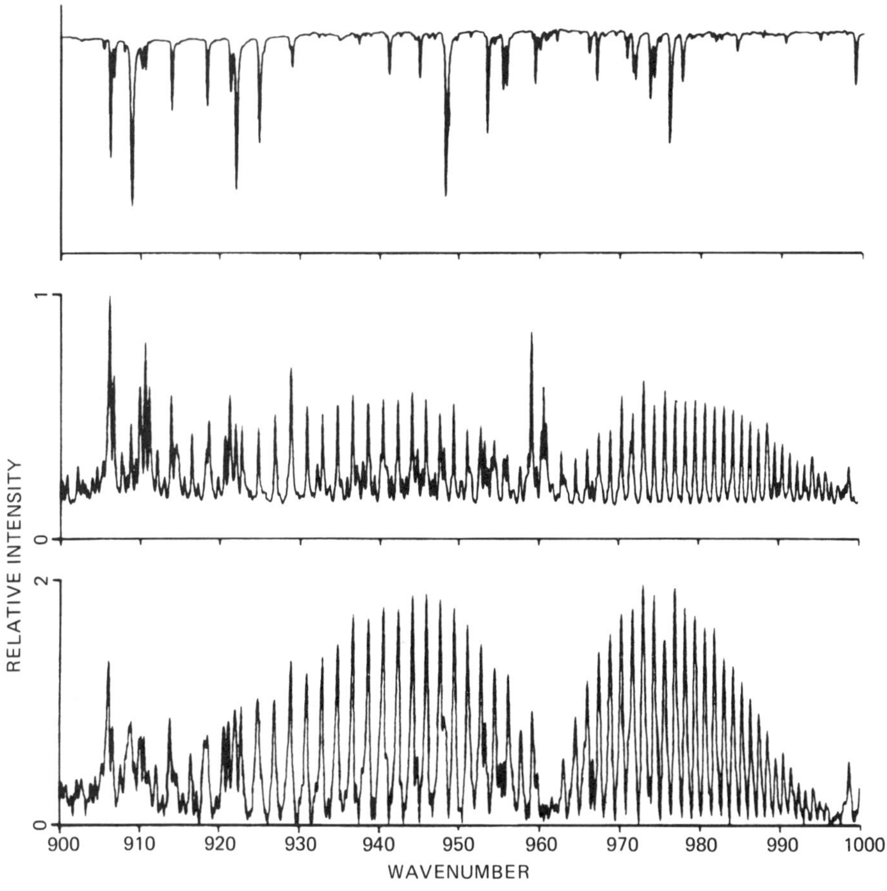

Fig. 12 Emission spectra of carbon dioxide from a carbon monoxide boiler (center trace) and a coal-burning power plant (lower trace) measured at distances of 600 and 800 m, respectively, using the E.P.A. ROSE system. The atmospheric transmission through a 560-m path is shown for comparison (upper trace) (from Herget and Brasher, 1980).

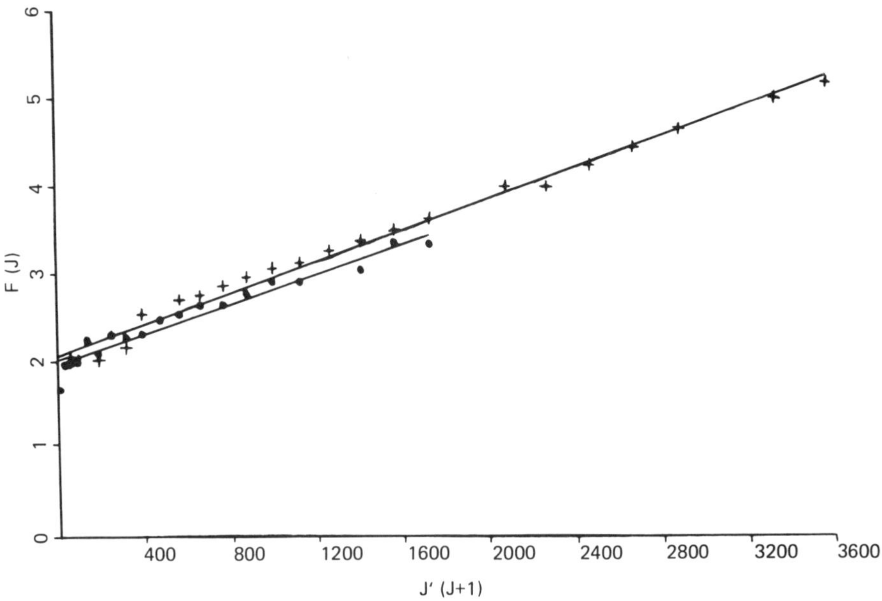

Fig. 13 Plot of $\ln[\bar{\nu}(m)^4/\gamma(m)I(m)]$ against $J'(J' + 1)$ for determination of the temperature of CO_2 from a carbon monoxide boiler plume from the spectrum shown in Fig. 12. +, *R* branch data; $T = 620$ K. ●, *P* branch data, $T = 645$ K. $\bar{\bar{T}} = 633$ K (from Herget and Brasher, 1980).

the infrared region is a recent development. Cheng and Bien (1981) reported studies of combustion gases in the laboratory in which a TDL spectrometer was used to probe the temperature profile of flames. They also discussed line choices for optimum sensitivity and for different optical depths. Measurements of this type have also been applied to the determination of the concentration of particular species in the flame at the same time.

An important advantage of the two-line method, which is also shared by the FT-IR methods of sensing temperatures, is that the measurement is based on a temperature-dependent physical property of the gas, so that the ratio of the intensities of two known lines is all the information needed to estimate the temperature of a sample. The theoretical ratio *F* of the line intensities was given by Hanson and Falcone (1978),

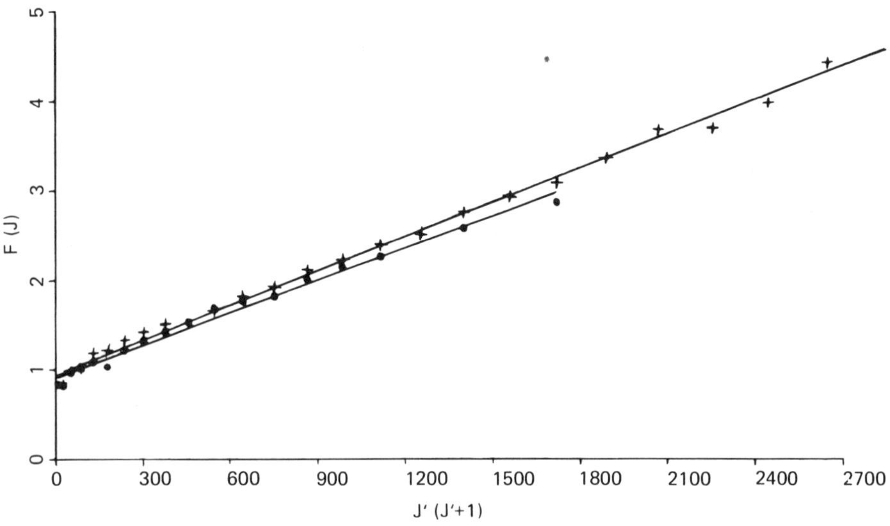

Fig. 14 Plot of $\ln[\bar{\nu}(m)^4/\gamma(m)I(m)]$ against $J'(J'+1)$ for determination of the temperature of the plume from a coal-burning power plant from the CO_2 hot band centered at 960 cm^{-1}. +, R branch data; $T = 430$ K. ●, P branch data; $T = 458$ K. Average, $\bar{T} = 444$ K (from Herget and Brasher, 1980).

$$F = \frac{\{(\nu + 1)(N_J/N_0)[\,|m|/(2J+1)]\}_1}{\{(\nu + 1)(N_J/N_0)[\,|m|/(2J+1)]\}_2} \tag{50}$$

where the values of ν and J in the numerator (line 1) denote the relevant vibrational and rotational quantum numbers for the vibrationally excited band and the values of ν and J in the denominator (line 2) are those for the line in the fundamental band. Line 2 may also be a hot band which is lower on the vibrational ladder than line 1. The population factor N_J/N_0 gives the fraction of molecules in the rotational level with quantum number J, which determines the ratio F for the two lines. This fraction is given by

$$(N_J/N_0) = \frac{(2J+1)}{Q(T)} \exp\left[\frac{-E_{\nu,J}(m)}{kT}\right] \tag{51}$$

where $E_{v,J}(m)$ is the combined rotational and vibrational energy of the energy state from which the line originates, and the quantum numbers v and J are also those of the initial state. $Q(T)$ is the partition function for vibration and rotation, as before.

Schoenung and Hanson (1981) have used this method to determine the temperature within a laminar flow flame and compared the result to thermocouple measurements made simultaneously. The temperature estimated by the two-line method, 1920 K, differed from that measured by the thermocouple by 80 K. The uncertainty in the spectroscopically measured temperature, however, was about 100 K.

Lowry et al. (1982) applied the two-line method to laminar flame temperature measurements using two pairs of lines: P(2), $v(1 \leftarrow 0)$ and R(4), $v(2 \leftarrow 1)$ at 2135.3 cm^{-1}, and the P(4), $v(1 \leftarrow 0)$ and R(2), $v(2 \leftarrow 1)$ pair at 2128.0 cm^{-1}. The line positions given are among those listed by Hanson and Falcone (1978). They also found good agreement between their spectroscopically determined temperature and both concurrent thermocouple measurements and theoretically calculated flame temperatures based on the proportions of fuel and oxidant. To measure line intensities, they fit a Voigt line shape to the measured line profile using calculated values of the collision width for the Voigt parameter a. Lowry et al. (1982) compare the temperature estimates obtained by the ratio of the intensities of the lines to those obtained by neglecting differences in linewidth and simply ratioing peak absorbances. They found that the latter method is a good approximation for these two pairs of lines, although the agreement with thermocouple measurements was slightly better when the ratio of line areas is calculated.

8 CONCLUSION

All of the techniques for estimating the temperature of hot gases from their infrared spectra assume that noise does not appreciably distort the relative intensities of each line in the spectrum. A few other assumptions have also been made, including the assumption of equipartition of energy within the rotational, vibrational, and translational degrees of freedom available to the absorbing molecules. We believe that this is usually a reasonable assumption for static samples, but that it is possible for the rotational and translational temperatures to differ for flowing samples or reacting systems. In addition, neighboring lines are assumed not to overlap appreciably. This assumption is valid provided that the resolution is high enough and the separation of the lines is sufficient. Numerical techniques have been developed to correct for the overlap when its effect is sufficient to affect the calculated temperature. These techniques would have to be applied, for example, for estimating the temperature of the sample from the spectrum of CO shown in Fig. 2.

The effects of noise, equipartition of energy and baseline errors are rarely important in practice, and often the only parameter which affects the accuracy to which temperatures can be estimated is the exponent for the temperature variation of the pressure broadening coefficient. The recent data of Varghese and Hanson (1981 and 1981a), Lowry et al. (1982), Sun and Griffiths (1981), Nakazawa and Tanaka (1982a), and Bouanich et al. (1983) suggest that even this limitation will shortly be overcome. When the temperature dependence of foreign-gas-broadening coefficients is fully understood, we believe that the temperature of hot isotropic gases will be measurable to an accuracy of ±2°C from their infrared spectra.

REFERENCES

Anderson, P. W. (1949). Pressure broadening in the microwave and infrared regions. *Phys. Rev.*, *76*: 647–661.

Anderson, R. J., and Griffiths, P. R. (1975). Errors in absorbance measurements in infrared Fourier transform spectrometry because of limited instrument resolution. *Analytical Chemistry*, *47*: 2339–2347.

Anderson, R. J., and Griffiths, P. R. (1977). Determination of rotational temperatures of diatomic molecules from absorption spectra measured at moderate resolution. *J. Quant. Spectrosc. Radiat. Trans.*, *17*: 393–401.

Bard, Y. (1974). *Nonlinear Parameter Estimation.* Academic Press, New York, pp. 233–238.

Birnbaum, G. (1967). Microwave pressure broadening and its application to intermolecular forces. *Adv. Chem. Phys.*, *12*: 487–548.

Bonamy, J., Robert, D., and Boulet, C. (1984). Simplified models for the temperature dependence of linewidths at elevated temperatures and applications to CO broadened by Ar and N_2. *J. Quant. Spectrosc. Radiat. Trans.*, *31*: 23–34.

Bouanich, J. P., and Brodbeck, C. (1973). Mesure des largeurs et des deplacements des raies de la bande O - 2 de CO. *J. Quant. Spectrosc. Radiat. Trans.*, *13*: 1–7.

Bouanich, J. P., and Farrenq, R. (1970). Largeurs des raies de la bande fondamentale de l'oxyde de carbone (CO) perturbe par l'argon et l'azote. *Comptes Rendus de l'Academie des Sciences, 271*: 507.

Bouanich, J. P., Farrenq, R., and Brodbeck, C. (1983). Direct measurements of N_2 broadened linewidths in the CO fundamental at low temperatures. *Can. J. Phys.*, *61*: 192–197.

Bouanich, J. P., and Haeusler, C. (1972). Linewidths of carbon monoxide self-broadening and broadened by argon and nitrogen. *J. Quant. Spectrosc. Radiat. Trans.*, *12*: 695–702.

Breene, R. G. (1981). *Theories of Spectral Line Shape.* John Wiley & Sons, New York, pp. 1–10.

Chackerian, C., Jr., Guelachvili, G., and Tipping, R. H. (1983). CO 1 - 0 band isotopic lines as intensity standards. *J. Quant. Spectrosc. Radiat. Trans.*, *30*: 107–112.

Cheng, W., and Bien, F. (1981). Temperature profile measurements using CO line absorption at different vibrational levels. *American Institute of Aeronautics and Astronautics J.*, *19*: 1071–1076.

Crane-Robinson, C., and Thompson, H. W. (1963). Pressure broadening studies on vibration-rotation bands, IV. *Proc. Roy. Soc. (London)*, *A272*: 453–466.

Darnton, L., and Margolis, J. S. (1973). The temperature dependence of the half widths of some self- and foreign-gas-broadened lines of methane. *J. Quant. Spectrosc. Radiat. Trans.*, *13*: 969–976.

Draegert, D. A., and Williams, D. (1968). Collisional broadening of CO absorption lines by foreign gases. *J. Opt. Soc. Am.*, *58*: 1399–1403.

Drayson, S. R. (1976). Rapid computation of the Voigt profile. *J. Quant. Spectrosc. Radiat. Trans.*, *16*: 611–614.

Eng, R. S., and Ku, R. T. (1982). High resolution linear laser absorption spectroscopy—review. *Spectros. Lett.*, *15*: 803–929.

Glenar, D. A. (1983). Review of progress in remote sensing by tunable diode laser heterodyne spectroscopy. *Tunable Diode Laser Development and Spectroscopy Applications* (W. Lo, ed.). *Proc. SPIE, 438*, pp. 125–136.

Griffiths, P. R. (1975a). *Chemical Infrared Fourier Transform Spectroscopy*, John Wiley & Sons, New York.

Griffiths, P. R. (1975b). Photometric precision in infrared spectra measured by Fourier transform spectroscopy. *Appl. Spectros., 29*: 11–14.

Griffiths, P. R. (1978). Fourier transform infrared spectrometry: theory and instrumentation, in *Transform Techniques in Chemistry* (P. R. Griffiths, ed.). Plenum Press, New York, pp. 109–139.

Griffiths, P. R. (1983). Fourier transform infrared spectrometry. *Science, 222*: 297–302.

Hanson, R. K. (1983). Tunable diode laser measurements in combustion gases, in *Tunable Diode Laser Development and Spectroscopy Applications* (W. Lo, ed.), *Proc. SPIE, 438*, pp. 75–83.

Hanson, R. K., and Falcone, P. K. (1978). Temperature measurement techniques for high temperature gases using a tunable diode laser. *Appl. Opt., 17*: 2477–2480.

Herget, W. F., and Brasher, J. D. (1979). Remote measurement of gaseous pollutant concentrations using a mobile Fourier transform interferometer system. *Appl. Opt., 18*: 3404–3420.

Herget, W. F., and Brasher, J. D. (1980). Remote Fourier transform infrared air pollution studies. *Opt. Eng., 19*: 508–514.

Herman, R., and Wallis, R. F. (1955). Influence of vibration-rotation interaction on line intensities in vibration-rotation bands of diatomic molecules. *J. Chem. Phys., 23*: 637–646.

Herzberg, G. (1950). *Spectra of Diatomic Molecules*. Van Nostrand, New York, p. 114.

Hindmarsh, W. R., and Farr, J. M. (1973). Collision broadening of spectral lines by neutral atoms. *Prog. Quant. Electron., 2*: 141–214.

Hirschfelder, J. O., Curtiss, C. F., and Bird, R. B. (1954). *Molecular Theory of Gases and Liquids*. John Wiley & Sons, New York, p. 22.

James, T. C., and Plyler, E. K. (1964). Linewidths in the 2-0 band of carbon monoxide broadened by nitrogen and hydrogen. *J. Chem. Phys.*, *40*: 221–223.

Kielkopf, J. F. (1973). New approximation to the Voigt function with applications to spectral-line profile analysis. *J. Opt. Soc. Am.*, *63*: 987–995.

Kim, H. (1970). Computer programming in physical chemistry laboratory. *J. Chem. Educat.*, *47*: 120–122.

Klim, A. (1981). A comparison of methods for the calculation of the Voigt profile. *J. Quant. Spectrosc. Radiat. Trans.*, *26*: 537–545.

Lambert, J. D. (1977). *Vibrational and Rotational Relaxation in Gases.* Oxford University Press, Oxford, pp. 1–12.

Low, M. J. D., and Clancy, K. F. (1967). Remote sensing and characterization of stack gases by infrared spectroscopy. An approach using multiple-scan interferometry. *Environ. Sci. Technol.*, *1*: 73–74.

Lowry, H. S. III, and Fisher, C. J. (1982). Line parameter measurements and calculations of CO broadened by nitrogen at elevated temperatures. *J. Quant. Spectrosc. Radiat. Trans.*, *27*: 585–591.

Lowry, H. S. III, Fisher, C. J., and Walker, H. C., Jr. (1982). High resolution tunable diode laser spectroscopy of carbon monoxide in gas mixtures and flames. Report for Arnold Engineering Development Center, Arnold Air Force Station, AEDC-TR-82-14, 64 pp.

Maitland, G. C., Rigby, M., Smith, E. B., Wakeham, W. A. (1981). *Intermolecular Forces, Their Origin and Determination.* Clarendon Press, Oxford.

Nakazawa, T., and Tanaka, M. (1982). Measurements of intensities and self- and foreign-gas broadened half-widths of spectral lines in the CO fundamental band. *J. Quant. Spectrosc. Radiat. Trans.*, *28*: 409–416.

Nakázawa, T., and Tanaka, M. (1982a). Intensities, half-widths, and shapes of spectral lines in the fundamental band of CO at low temperatures. *J. Quant. Spectrosc. Radiat. Trans.*, *28*: 471–480.

Nielsen, J. R., Thornton, V., and Dale, E. B. (1944). The absorption laws for gases in the infra-red. *Rev. Mod. Phys.*, *16*: 307–323.

Noll, R. J., and Pires, A. (1980). A new nonlinear least squares algorithm for Voigt spectral lines. *Appl. Spectrosc.*, *34*: 351–360.

Partin, D. L. (1983). Heterojunction stripe geometry lead salt diodes grown by molecular beam epitaxy. *Tunable Diode Laser Development and Spectroscopy Applications* (W. Lo, ed.). *Proc. SPIE, 438*, pp. 17–20.

Penner, S. S. (1959). *Quantitative Molecular Spectroscopy and Gas Emissivities*, Addison-Wesley Publishing Company, Inc., Reading, Pa., pp. 457–460.

Rinsland, C. P., Goldman, A., Murcray, F. J., Murcray, D. G., Smith, M. A. H., Seals, R. K. Jr., Larsen, J. C., and Rinsland, P. L. (1983). Stratospheric temperature measurement profile from balloon-borne measurements of the 10.4 μm band of CO_2. *J. Quant. Spectrosc. Radiat. Trans.*, *30*: 327–334.

Rolland, C., Reid, J., Garside, B. K., Morrison, H. D., and Jessop, P. E. (1984). Investigation of a CW optically pumped 12 μm NH_3 laser using a tunable diode laser. *Appl. Opt.*, *23*: 87–93.

Schoenung, S. M., and Hanson, R. K. (1981). CO and temperature measurements in a flat flame by laser absorption spectroscopy and probe techniques. *Combustion Sci. Technol.*, *24*: 227–237.

Sell, J. A. (1980). Temperature dependence of the absorption coefficient and halfwidth of the P(6) line of carbon monoxide. *J. Quant. Spectrosc. Radiat. Trans.*, *23*: 595–598.

Sun, N. N-P., and Griffiths, P. R. (1981). Temperature dependence of the self-broadening coefficients for the fundamental band of carbon monoxide. *Appl. Opt.*, *20*: 1691–1695.

Sun, J. N-P., and Griffiths, P. R. (1981a). Nitrogen-broadening coefficient of vibration-rotation lines of carbon monoxide. *Appl. Opt.*, *20*: 2332–2334.

Sun, J. N-P., Olson, M. L., Grieble, D. L., and Griffiths, P. R. (1980). Rapid scanning computer controlled tunable diode laser spectrometry. *Appl. Opt.*, *19*: 2762–2766.

Tejwani, G. D. T. (1972). Half-width computations for CO broadened lines. *J. Quant. Spectrosc. Radiat. Trans.*, *12*: 123–128.

Tsao, C. J., and Curnutte, B. (1962). Line widths of pressure broadened spectral lines. *J. Quant. Spectrosc. Radiat. Trans.*, *2*: 41–91.

Varanasi, P. (1981). Comment on the intensity and the temperature dependence of the nitrogen-broadened half-width of the P(6) line in the CO fundamental. *J. Quant. Spectrosc. Radiat. Trans.*, *25*: 187–188.

Varanasi, P., and Sarangi, S. (1975). Measurements of intensities and nitrogen-broadened linewidths in the CO fundamental at low temperatures. *J. Quant. Spectrosc. Radiat. Trans.*, *15*: 473–481.

Varghese, P. L., and Hanson, R. K. (1980). Tunable infrared diode measurements of line strengths and collision widths of $^{12}C^{16}O$ at room temperature. *J. Quant. Spectrosc. Radiat. Trans.*, *24*: 479–489.

Varghese, P. L., and Hanson, R. K. (1981). Collision width measurements of CO in combustion gases using a tunable diode laser. *J. Quant. Spectrosc. Radiat. Trans.*, *26*: 339–347.

Varghese, P. L., and Hanson, R. K. (1981a). Room temperature measurements of collision widths of CO lines broadened by H_2O. *J. Molec. Spectrosc.*, *88*: 234–235.

Wang, J. Y. (1976). Laser absorption methods for simultaneous determination of temperature and species concentrations through a cross section of a radiating flow. *Appl. Opt.*, *15*: 768–773.

Whiting, E. H. (1968). An empirical approximation to the Voigt profile. *J. Quant. Spectrosc. Radiat. Trans.*, *8*: 1379–1384.

4

Flare Efficiency Monitoring by Remote Infrared Sensing

ROBERT L. SPELLICY *OptiMetrics, Inc., Las Cruces, New Mexico*

MERLE J. PERSKY* *OptiMetrics, Inc., Bedford, Massachusetts*

Current affiliation: Lincoln Laboratory, Massachusetts Institute of Technology, Lexington, Massachusetts

1 INTRODUCTION

Flares are used by a number of industries for the disposal of waste gases by combustion. One study estimates that a total of 16 million tons of gas are burned per year in the United States (Joseph et al., 1983). Since the volume of gases burned is so large, both the government and the associated industries have a vital interest in ensuring that the design and operation of flare systems minimize the production of undesirable by-products. An effective means by which flare efficiency can be monitored is therefore essential. To meet this need, the authors have developed methods for using *remote* infrared spectroscopy to measure the concentration of combustion by-products of industrial flares and thereby monitor the combustion efficiency of flare systems. To test these methods, species concentrations from a small-scale industrial flare were measured using two methods:

1. Point sampling of combustion products using extractive probes and
2. Remote sensing of combustion products using the developed infrared spectroscopic techniques

Comparison of the data taken by both methods indicates the basic validity and applicability of the Fourier transform infrared (FT-IR) approach to efficiency monitoring. The ATLES (Atmospheric Transmission for Line Emission Sources) radiation code was used to model observed CO_2 radiation while line-by-line codes were used to evaluate CO concentrations. Because the validity of the ATLES code was somewhat uncertain, the accuracy and range of validity of this code were also tested using a set of controlled laboratory measurements.

1.1 Measurement of Flare Combustion By-Products

Both measurement approaches, point sampling and remote sensing, have strengths and weaknesses. The extractive probe provides localized data on a large number of exhaust species in a very direct manner; however, the local nature of the probe measurement makes it vulnerable to wind effects unless compensation is provided by including a tracer gas or by using the *ratios* of combustion products rather than absolute values. That is, variations in concentrations observed with a probe could be caused by the wind, moving the combustion products relative to the probe, or they could be simply attributable to spatial variations in the plume itself. In addition, because the probe is a direct measurement tool, the apparatus must be at the flare itself, possibly causing interference with plant operation.

Remote optical sensing offers great promise for flare efficiency monitoring because it need not interfere with plant operations. The remote sensing concept can be used because the exhaust products employed in the efficiency calculation are infrared-active, meaning that each has its own unique radiant emission (or signature) in the infrared region of the spectrum. This signature, because it is related to the concentration of the product, allows infrared spectroscopy to be used to determine flare efficiency by measuring these radiant emissions remotely and nonintrusively. The infrared method also may provide a more representative measure than a localized probe, because spatial averages rather than specific point data can be obtained.

1.2 Validation of Remote Sensing Concepts

To test the validity of the remote sensing concepts, we collected infrared spectroscopic data on a small-scale, industrial flare using the EPA ROSE (Remote Optical Sensing of Emissions) system. At the same time, others took extractive probe measurements on the flare for point sampling of combustion products. The probe measurements provided localized data on gas temperatures and on the concentrations of such combustion products as CO, CO_2, hydrocarbons, and particulates. Since these probe measurements represent the current state of the art in flare emission monitoring, they were used to compare with the FT-IR results to estimate the basic validity of the infrared spectroscopic approach. For the purpose of accurate comparisons, the majority of the remote infrared data were taken at the same position and at the same time as the extractive probe data were taken.

A further comparison of FT-IR and probe data was performed using a set of five unknown calibration gases whose identities and concentrations were determined both by the extractive probe analysis methods and by absorption measurements using the ROSE FT-IR. These comparisons were used to determine the inherent agreement between the two measurement methods independent of flare or atmospheric influences.

To understand the results of these comparisons, we should consider what is meant by flare efficiency. There are a number of different definitions of efficiency, some of which require the use of trace compounds added to the flare fuel; others require the monitoring of plant activity in one way or another. Extractive probe techniques use the term "combustion efficiency," which is based on the oxidation of flare emissions to carbon dioxide. It is defined as (McDaniel, 1983)

$$\text{EFF} \equiv \frac{[CO_2]}{[CO_2] + [CO] + [THC] + [Soot]} \tag{1}$$

where $[CO_2]$ is parts per million by volume of carbon dioxide, [CO] parts per million by volume of carbon monoxide, [THC] parts per million by volume of total hydrocarbon as methane, and [Soot] parts per million by volume of soot as carbon.

The same definition was adopted for our remote sensing technique, except the soot term was deleted because soot could not be monitored directly with the available infrared equipment. Fortunately, few of the tests studied had significant soot loadings. Furthermore, the shallow optical depths of the flare studied were insufficient for measuring hydrocarbon radiant emissions. This necessitated using an approximation to Eq. (1), depending only on carbon monoxide (CO) and carbon dioxide (CO_2) as

$$\text{EFF} \approx \frac{[CO_2]}{[CO_2] + [CO]} \tag{2}$$

Since, in general, the total hydrocarbon concentration is small compared with that of CO and CO_2 at high efficiencies, this approximation is appropriate for evaluation of flare operation in the high efficiency regime encountered with most industrial flares.

2 FIELD MEASUREMENTS

2.1 The Remote Optical Sensing of Emissions (ROSE) System

A versatile measurement tool, the ROSE system contains an FT-IR spectrometer as well as computer-based data collecting and processing equipment (Herget and Brasher, 1980; Herget, 1982). This same system was used earlier to take the data analyzed in Chap. 1. The essential elements of the this system are

1. A Nicolet Model 7199 Fourier transform interferometer with an f/5, 0.6-m Dall–Kirkham telescope, and
2. A Nicolet Model 1180 digital computer and data acquisition system

A schematic of the ROSE optical system is shown in Fig. 1, while Table 1 provides a summary of the major parameters of the system as configured for the flare tests.

The ROSE system is capable of taking measurements over the entire spectral region from 600 to 6000 wavenumbers using a sandwich detector; however, to obtain most accurate measurements, a single detector was used to limit the spectral response to the region of most importance—from 1800 to 6000 wavenumbers. Spectral resolution, another adjustable system parameter, was fixed at 1/4 wavenumber for these tests. This value was the best compromise between signal-to-noise ratio, which *decreases* with increasing resolution, and gas detectability, which *increases* with increasing resolution.

The area observed within the flare is dictated primarily by the 0.6-m size of the receiver telescope, since the acceptance angle of the ROSE system is less than 1 mrad. Consequently, the sampled area was also fixed for all the measurements.

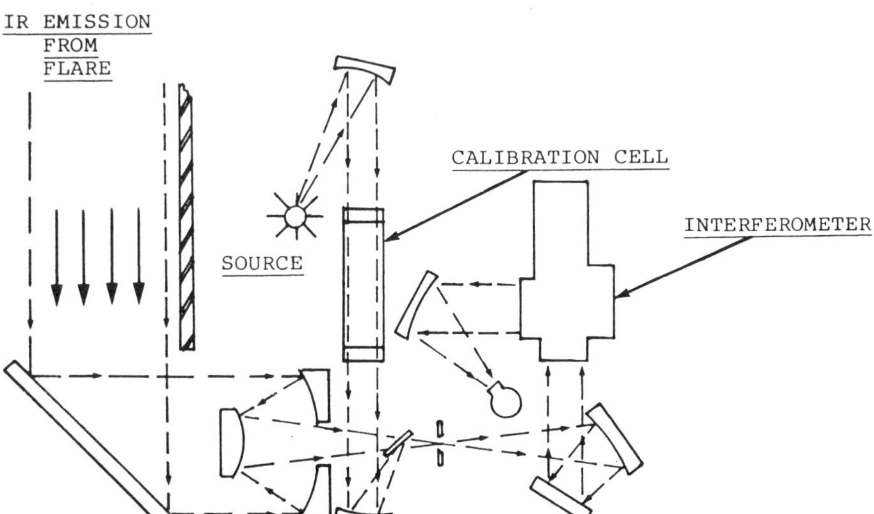

Fig. 1 ROSE optical system.

TABLE 1 ROSE System Operating Parameters for Flare Measurements

Parameter	Value
Spectral range	
primary measurements	1800–2500 cm^{-1}
additional measurements	800–1800 cm^{-1}
Spectral resolution	1/4 cm^{-1}
Field of view (at flare)	0.60 m
Observation times[a]	4, 40 or 200 sec

[a]Sample-to-sample variable.

2.2 The Extractive Probe

An extractive probe was used simultaneously with the ROSE system in order to obtain independent measurements of plume temperature as well as CO, CO_2, NO_x, SO_2, and total hydrocarbon concentrations. Guy wires and a crane were used to adjust the probe position.

Three basic types of samples were collected (McDaniel, 1983):

1. Continuous samples of CO, CO_2, NO_x, and SO_2, recorded every 10 sec with an instrumental response time of 20 sec
2. Instantaneous samples for gas chromatographic analysis of individual hydrocarbon species, collected every 5 or 6 min with a 20-sec sampling window; and
3. Integrated samples, collected into bags over time periods of 5–15 min and subsequently analyzed by gas chromatography

Continuous samples, as in method (1) above, were used to provide the point sampled concentrations used for comparison with the results of the fitting methods derived here.

2.3 Data Collection

The ROSE system collects spectra using the radiation emitted by the hot gases in the flare plume. However, these data must be corrected for ambient radiation levels and calibrated in absolute radiance units before they can be analyzed to identify the species present or their concentrations. A full measurement sequence therefore consists of collecting several different types of data:

1. Flare emissions data containing information on the combustion products
2. Background data quantifying the ambient radiation levels, and
3. Calibration data defining inherent system response characteristics

Procedures used for the collection of each of these types of data are given below, after which a comprehensive discussion is presented of the procedures required to combine these data sets into absolute, calibrated results.

Flare Emissions

For the majority of the emission measurements, the FT-IR field of view was positioned coincident with the extractive probe, typically several feet above the end of the combustion zone. In this area, the sample was estimated to be most representative of the final species concentrations. FT-IR data were also taken of sample areas 2, 6, and 16 ft above the flare tip as well as within the visible flame. The purpose of these latter measurements was to determine the change in species concentrations with height and to substantiate the assumption that the optimum flare location for spectroscopic measurement of efficiency is above the combustion zone rather than within it.

During collection of flare emission data, the wind caused the flame to move in and out of the field of view of the interferometer. Since this could cause temporal fluctuations in the data, which adversely affected the spectrum, several different averaging periods were used to insure collection of more accurate data. Because of this wind-induced bending of the flame, it was also necessary to move the extractive probe in order to keep it within the plume. To ensure that the region being sampled was in the flowfield above the reaction zone, the probe had to be moved from run to run and frequently during a given run as well. Since the FT-IR data were to be compared with those of the probe, if the probe was relocated, the steering mirror on the ROSE system was used to maintain the FT-IR field of view coincident with the probe. In this way, common sampling areas were always being used by the two systems.

Optical Backgrounds

Background data were required to establish the ambient radiation levels, ambient species concentrations, and ambient atmospheric-constituent levels present at the measurement site. The first two of these are required to correct the flare data for radiation sources or ambient gas emissions/absorptions within the system field of view. The last is required to quantify the propagation characteristics of the atmosphere between the flare and the ROSE system, allowing for interpretation of the observed spectrum. To collect these background data, emission scans identical to those taken on the flare itself were collected just before or just after a flare run, using the same fields of view as much as possible.

Calibration Data

The spectral features of the infrared spectrum identify the particular constituents present in the flare. The intensity of these features then dictates the concentrations of the species giving rise to these spectral features. However, to quantify these concentrations, an absolute calibration of the instrument is essential. Absolute calibration allows for extraction of the actual power per unit solid angle per unit wavelength arising from the emitting gases. To achieve such calibration, additional emissions data are collected with the interferometer system, but here a blackbody source is used for which the total power output with wavelength and solid angle is a calculable quantity.

To collect data of this type for the ROSE system, a Barnes blackbody was used in conjunction with a 0.6-m diameter Dall-Kirkham transmitter/telescope. This combination, used at a distance of 59 ft from the van, supplied a radiation field of known intensity, filling the field of view of the ROSE system. Using data collected in this configuration, one could correct the data for instrumental effects and extract absolutely calibrated data.

Calibration of FT-IR

The output of the FT-IR, known as an interferogram, is not directly usable for analysis. Although this interferogram contains all the available spectral information, it must be Fourier transformed before it is recognizable as an infrared spectrum. Once this is done, the resultant spectrum then must be background corrected and calibrated before it can be used to extract species concentrations or flare efficiencies. The procedures used to perform these tasks are outlined below in the form of five data reduction steps.

Step 1: generate low resolution spectra. A mathematical process known as a fast Fourier transform (FFT) was applied to all interferograms (flare emission, optical backgrounds, and calibrations) to generate infrared spectra. Since a low resolution (2 wavenumbers) FFT requires only a few seconds of computer time, whereas a high resolution (1/4 wavenumber) transform takes up to 30 min, the first step was to perform low resolution transforms on all of the collected data in order to provide hard copy spectra suitable for data selection. Approximately 100 low resolution transforms were obtained and plotted in this way. These plots are adequate only for evaluation of the potential usefulness of each spectrum in subsequent flare efficiency analysis; the needs of the analysis can be completely satisfied only using full high resolution data.

Step 2: perform data quality evaluation. The low resolution plots were reviewed and categorized as to relative usefulness before further processing. The criteria used in this initial evaluation were signal-to-noise ratio, the absence of signal saturation, and the indication of observable gas emissions. A further sorting was performed to create a matrix of spectra which would best represent the low, medium, and high flow rates of propylene, steam, and nitrogen, as well as to represent the several spatial sampling positions. Background and calibration data appropriate to those specific flare spectra were also identified.

Step 3: generate full resolution spectra. High resolution (1/4 wavenumber) spectra were then generated for the data indicated in Table 2. Each resulting spectrum was transferred to disks for permanent storage and for subsequent analysis.

Step 4: spectrometer calibration. The spectrometer calibration relates the spectral amplitudes in the spectra to the intensity of the infrared energy received by the system at each wavenumber. In general:

$$S = RP \tag{3}$$

where S is the observed amplitude, R the response of the total FT-IR system, and P the power received by the system.

TABLE 2 Flare Test Parameters for FT-IR Samples

OMI No.	Test No.	Day	Local time	Sampling point	Sample time (sec)	Propylene flow (lb/hr)	Nitrogen flow (lb/hr)	Steam flow (lb/hr)	Exit velocity (ft/min)
7	2	6/17	17:13:08	(1)	400	3078	-	1564	2475
12	3	6/18	10:03:33	36	200	3027	-	1355	2432
76	4	6/22	21:45:30	47	4	1875	-	-	1509
23	5	6/18	17:00:52	41	136	991	-	1543	795
25	5	6/18	17:14:06	35	400	991	-	1543	795
95	11a	6/24	07:32:13	41	4	612	2489	-	3520
124	16a	6/24	10:59:29	31	4	329	1182	-	1907
118	16b	6/24	11:03:24	(2)	40	313	897	-	1344
119	16b	6/24	11:07:38	(2)	200	313	897	-	1344
42	17	6/21	12:49:55	25	40	162	-	150	131
44	17	6/21	12:56:54	25	4	162	-	150	131
58	23	6/21	19:19:30	29	4	0.451	2.13	-	2.63
50	50	6/21	14:10:00	25	56	162	-	498	130

54	51	6/21 18:05:23	29	4	305	1222	256	1733
61	52	6/21 19:59:10	19	200	0.452	2.14	201	2.96
68	53	6/22 16:26:02	18	40	0.226	1.41	201	1.90
71	54a	6/22 17:19:59	18	40	0.226	1.41	-	1.90
73	54b	6/22 17:30:32	19	200	0.226	1.41	-	1.90
88	55	6/23 14:04:34	26	200	164	-	1125	132
89	55	6/23 14:06:49	26	200	164	-	1125	132
90	56	6/23 14:49:38	25	4	163	-	562	131
109	57a	6/24 09:21:06	42	40	629	2663	497	3749
129	59a	6/24 21:50:03	(3)	200	345	2361	-	3152
132	59b	6/24 21:55:16	37	200	350	1942	-	2645
137	60d	6/24 23:13:15	(4)	4	212	1325	-	1781
139	60d	6/24 23:24:38	(5)	4	212	1325	-	1781

Note: Sample area for spectrometer if not at probe: (1), near exit; (2), 6 ft above flare tip; (3), top of flame; (4), top of flame; (5), 2 ft above flare tip.
Source: McDaniel, 1983.

If the system is used to collect emissions data while viewing a blackbody calibration source, the signal received will be a combination of the blackbody energy (BB), radiation arising from backgrounds within the field of view (BKG), and emissions arising from the ambient temperature optics of the interferometer itself (SPEC). Consequently,

$$S(BB + BKG + SPEC) = R[P(BB) + P(BKG) + P(SPEC)] \tag{4}$$

By independently measuring the background signal, which also contains the spectrometer self-emission signals, one obtains

$$S(BKG + SPEC) = R[P(BKG) + P(SPEC)] \tag{5}$$

If this is subtracted from Eq. (4), the instrument response may be evaluated as

$$R = S(BB)/P(BB) \tag{6}$$

Since $P(BB)$, the power arising from the blackbody, is defined by the Planck radiation function, a measurement of the background signal and the observed blackbody signal completely defines the system response [Eq. (4)].

Step 5: calibration of observed flare data. Data collected during a flare test has the same form as that of Eq. (4). That is,

$$S(\text{Flare} + BKG + SPEC) = R[P(\text{Flare}) + P(BKG) + P(SPEC)] \tag{7}$$

If an independent measurement of background emissions is taken for the flare measurement configuration and this is subtracted from the raw data, the calibration factor can be applied to give P(Flare) directly. We can recall that in Chap. 2, Eq. (12), which assumed spectrometer self-emission to be small, also made an explicit division between a background contribution from the region around the flame or plume and a contribution of radiation from behind the hot gas and passing through it. When the emitting gas fills only a small fraction of the field of view, or when it is optically thin, the background contribution in Eq. (7) can be measured by simply observing the background without the source present. That is,

$$P(\text{Flare}) = \frac{1}{R}[S(\text{Flare} + BKG + SPEC) - S(BKG + SPEC)] \tag{8}$$

This then may be put into radiance or irradiance units by dividing by the appropriate receiver areas or solid angles.

3 REDUCTION OF FLARE DATA

The specific analytical procedures developed for reduction of flare data are outlined in Fig. 2. Assuming that one begins with calibrated field data, temperatures and concentrations are evaluated using the six-step procedure shown.

Step 1. The first step is to examine the spectral data, looking at empirical and theoretical line positions of the several gases of interest to determine the species which are present in sufficient concentrations to yield usable signatures. With carbon dioxide, for example, the process is straightforward, since there is a strong band (the ν_3 fundamental) in the region from 4 to 5 μm (2000–2500 cm^{-1}) which has been well characterized in the past. Similar bands exist for all the major species.

Step 2. After locating the major molecular bands of interest, the second step is to determine an *initial value* for the exhaust temperature. This is done using the band envelope of the CO fundamental band between 2000 and 2140 cm^{-1} by making a semilogarithmic plot of

$$\frac{N_{\bar{\nu}}}{\bar{\nu}^4 |m|} \quad \text{versus } J' (J' + 1) \tag{9}$$

where $N_{\bar{\nu}}$ is the radiance of a given CO line, $\bar{\nu}$ the wavenumber of that line, J' the quantum number of the initial (upper) state of the transition, and $|m| = J' + 1$ (for the P branch used).

In the absence of gradient or large optical depth effects (see Chap. 3, especially Sec. 7.1), the slope of this plot is (Herzberg, 1960)

$$\frac{B'hc}{kT} \tag{10}$$

where B' is the spectroscopic rotational constant (in the upper state) for the molecule under observation, h Planck's constant, c the speed of light, k Boltzmann's constant, and T the desired temperature (Kelvin).

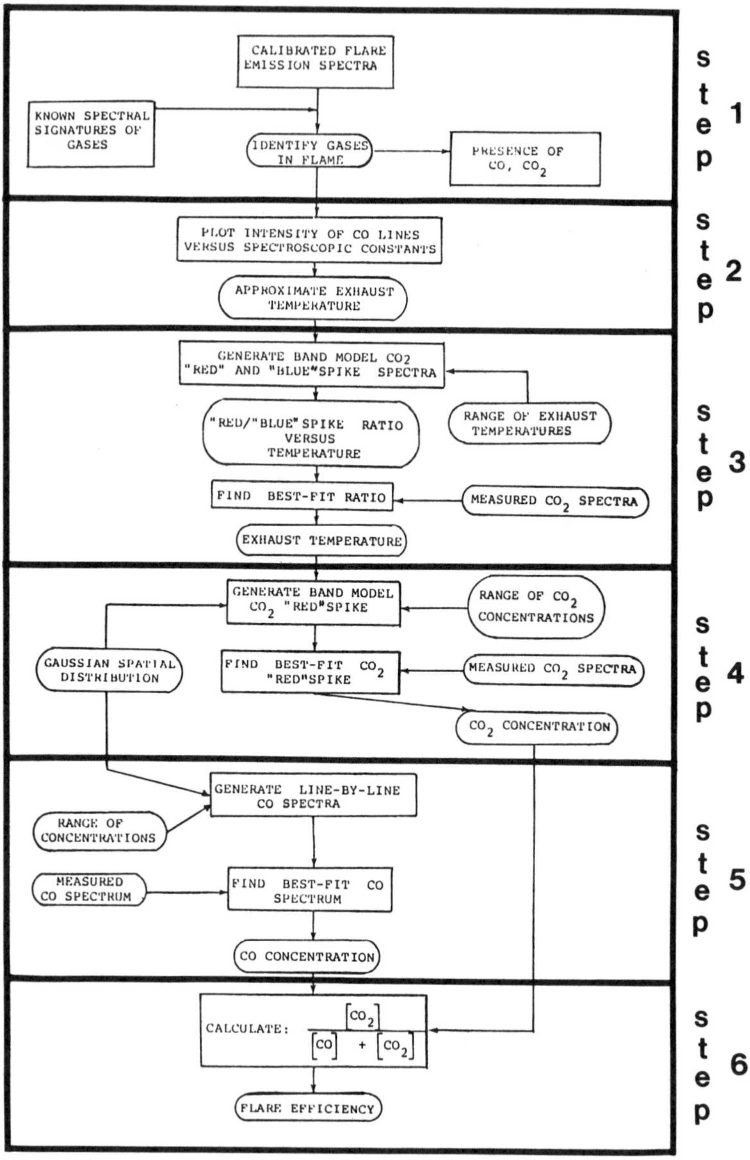

Fig. 2 Analysis procedure developed for flare efficiency evaluation.

In actual scenarios where large gradients or optical depths are present, care must be taken in assuming that this relationship is valid.

Step 3. The initial temperature estimate is then used as a starting point for the generation of a matrix of computer-synthesized, low resolution, CO_2 spectra used to evaluate a refined plume temperature. The CO_2 calculations performed here were based on a band model code called ATLES (Young, 1977) to produce a set of low resolution spectra in the so-called red- and blue-spike regions of the ν_3-fundamental band. As shown in Fig. 3, the red spike extends from approximately 2200 to 2325 cm^{-1} while the blue spike extends from 2375 to 2400 cm^{-1}. As discussed in Chap. 1, these spikes are caused by CO_2 atmospheric absorption, which absorbs the center of the CO_2 plume emissions before they reach the interferometer, leaving two spikes in the wings of the band.

Since the flare plume is not homogeneous, calculation of these infrared emissions requires some measured or assumed spatial distribution for both the temperature and the species concentrations in the plume. Without any measured values for this distribution, a Gaussian profile such as that shown in Fig. 4 was assumed for the present analysis. To test the sensitivity of the results to the profile chosen, a comparison was made between two profiles which had a factor of two difference in the speed with which their core values drop to ambient (roll-off). As indicated in Fig. 5, the band model calculations were not significantly affected by this roll-off when the core temperature was held constant.

It was observed, however, that the *ratio* of the peak intensities of the red to blue spikes is (1) a strong function of peak temperature, as illustrated in Fig. 6, and (2) independent of the concentration, as indicated in Fig. 7. Consequently, the next step in the analysis procedure is to match computer-generated CO_2 spectra to the measured spectrum for a given test, using the ratio of the amplitudes of the red and blue spikes to obtain a refined temperature estimate. This refined temperature is then used in the next step of the process to determine CO_2 concentrations.

Step 4. To determine CO_2 concentrations, the ATLES code is used again to produce a set of comparison spectra, but in this case, concentration instead of temperature is varied. Examples given in Fig. 8 illustrate the uniform increase of intensity for both the red and blue spikes with concentration. Given the near-

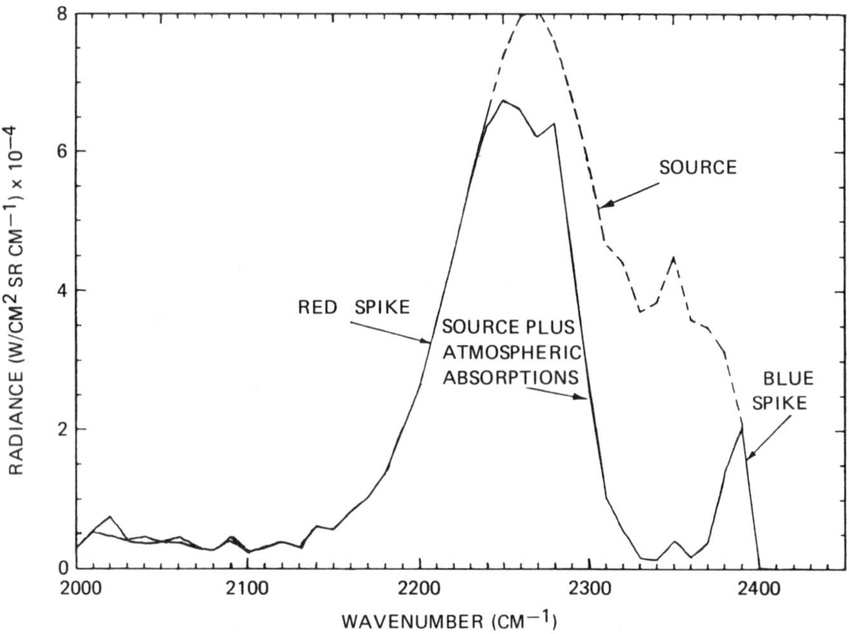

Fig. 3 Creation of red and blue spikes of CO_2.

ly continuous absorption of the red spike, it is used exclusively for determining a best-fit match with the measured CO_2 spectra because its peak intensity is relatively insensitive to instrumental resolution. Once a best-fit match is found, the CO_2 concentration is determined.

Step 5. To evaluate carbon monoxide concentrations in the plume, a similar best-fit procedure is used. In this procedure however, a high resolution comparison is necessary because the CO emissions arise from relatively isolated emission *lines*, as shown in Fig. 9, instead of continuous bands as in the CO_2 red spike. Even though the resolution required is different, the rest of the fitting procedure is essentially the same as that used for the CO_2. The temperature and species distributions determined from the CO_2 fits are used directly in the CO calculations. The CO concentration is then varied to find a best match to observed data.

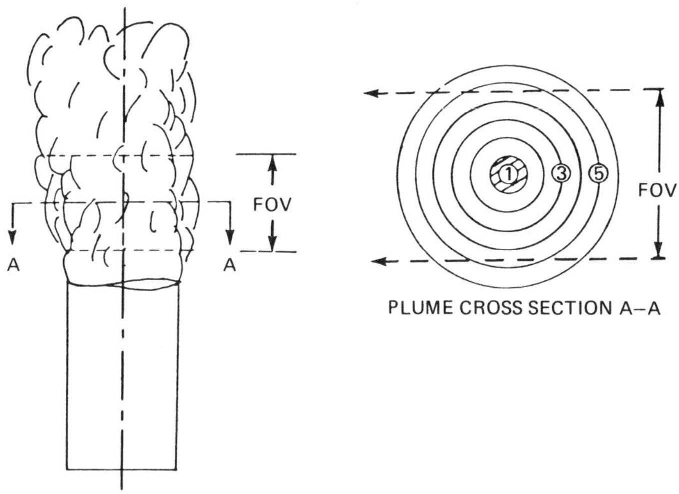

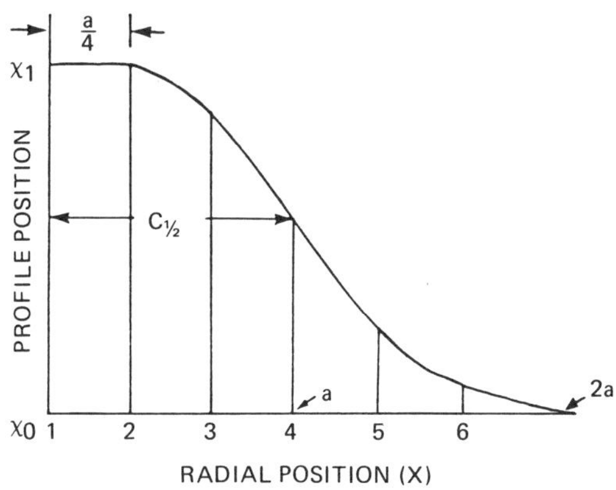

Fig. 4 Gaussian spatial distribution. The mole fraction χ is given by $\chi(x) = \chi_0 + \chi_1 \exp[-\ln 2(x - x_0)^2/a^2]$. χ_0, ambient level; χ_1, peak level above ambient; $x_0 = 0.812$ m; $a = 0.406$ m.

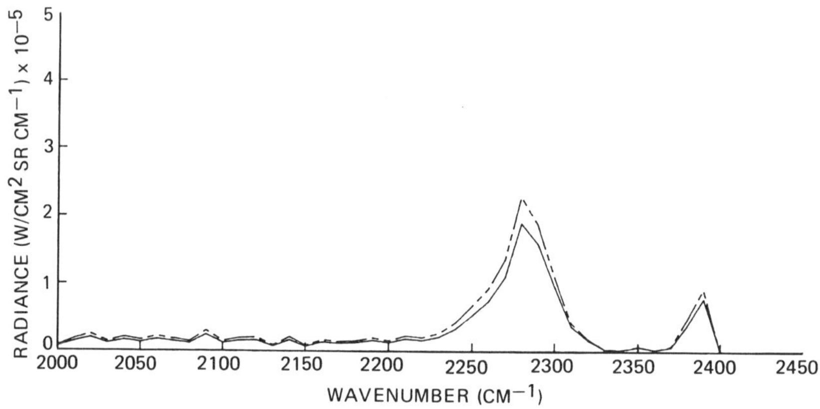

Fig. 5 Radiance variations caused by altering Gaussian profile roll-off (lower curve rolls off twice as rapidly as upper); test 55, OMI No. 89.

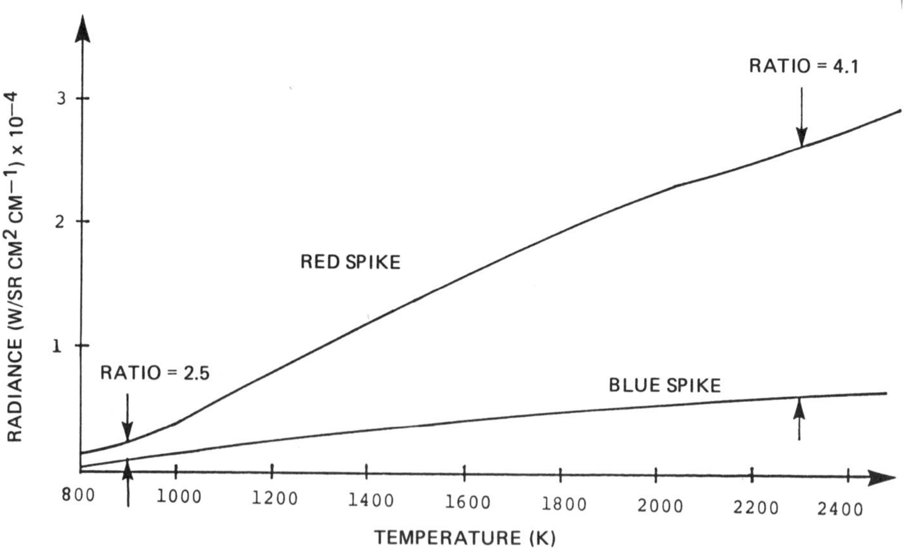

Fig. 6 Red and blue spike intensities versus temperature.

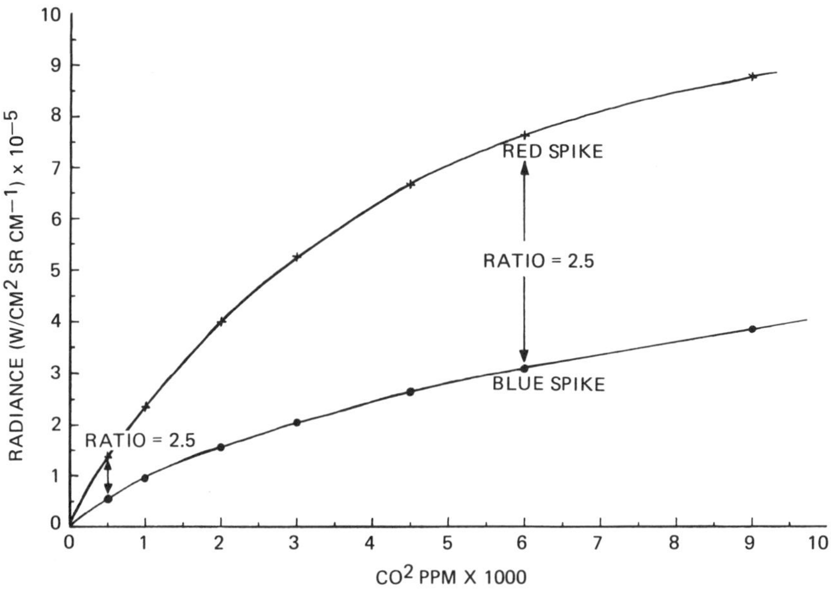

Fig. 7 Variation in peak red spike and blue spike radiance with CO_2 concentration at 800 K.

To perform the CO spectrum calculations, an algorithm called FASCOD1 (Clough et al., 1981) is used to effectively compute the contributions of each of the individual CO lines and then to combine them into a single spectrum. The code has been extensively verified and has several characteristics rendering it well suited to this particular task:

1. It computes the plume radiance as well as the transmittance of the intervening atmosphere and does so at a resolution that is high compared to typical instrumental capabilities.
2. It assumes local thermodynamic equilibrium characterized by temperature and pressure.
3. It allows for inclusion of an instrument scanning function so that the resolution of the calculations can be matched to that of the measurements for optimum spectra comparisons.

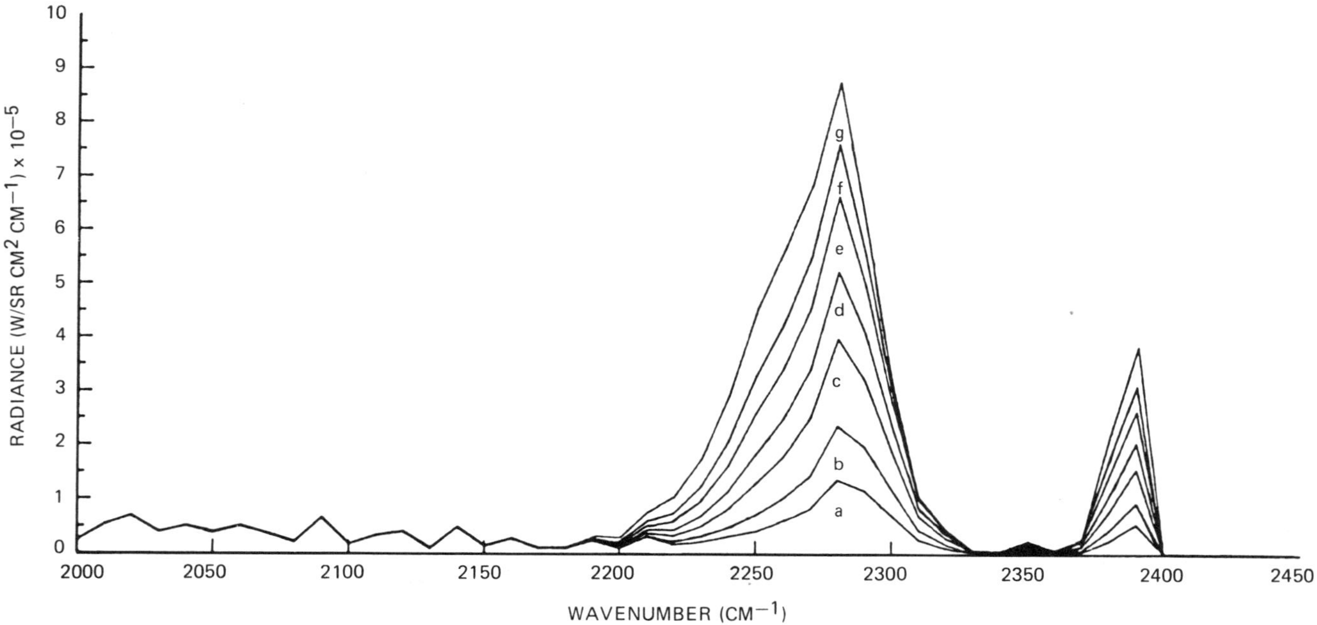

Fig. 8 Families of curves for a flare temperature of 800 K and concentrations of (a) 500, (b) 1600, (c) 2000, (d) 3000, (e) 4500, (f) 6000, and (g) 9000 ppm of CO_2.

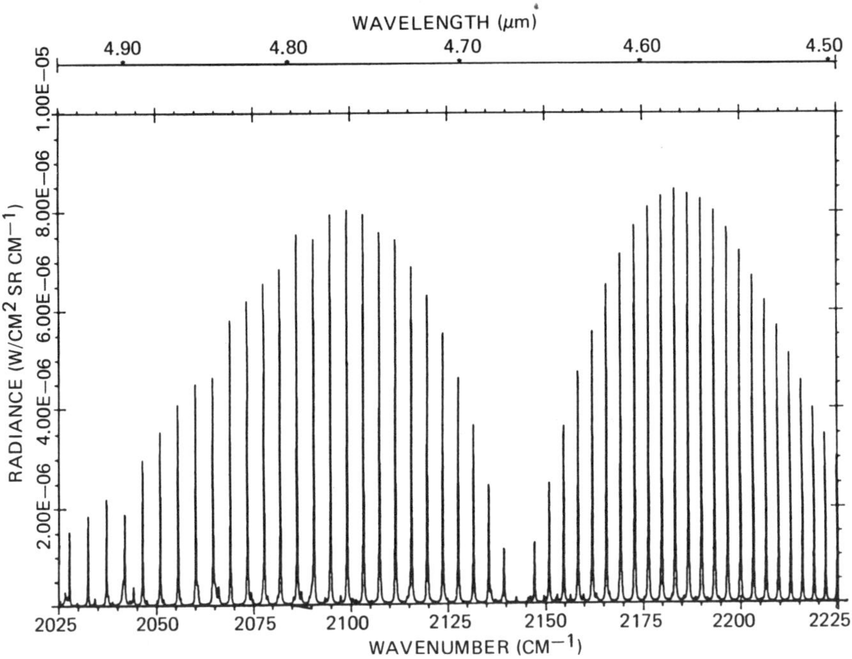

Fig. 9 Predicted CO spectrum.

Because of water vapor in the plume and in the intervening atmosphere, some potential exists for emission line interference in the CO region. To minimize the impact of this interference, the calculation is performed for some 50 lines in both the P and R branches. A best-fit match to all the lines is then used to evaluate the CO concentration for that test.

Step 6. The final step in the data analysis process is to calculate flare efficiency from Eq. (2):

$$\mathrm{EFF} \approx \frac{[\mathrm{CO_2}]}{[\mathrm{CO}] + [\mathrm{CO_2}]} \tag{11}$$

The validity of this relationship at high efficiency levels is indicated in Figs. 10a, 10b, 10c, and 10d. The data plotted in these figures are the results of the continuous extractive probe

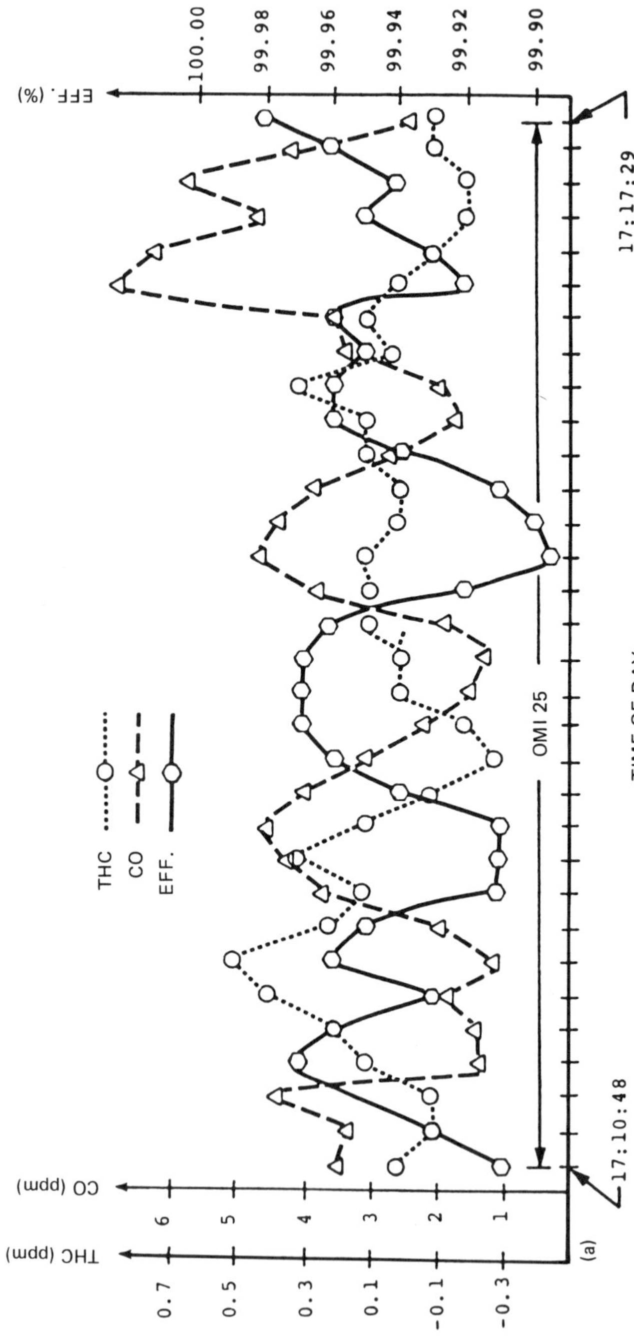

Fig. 10a Continuous probe sampling results, test 50.

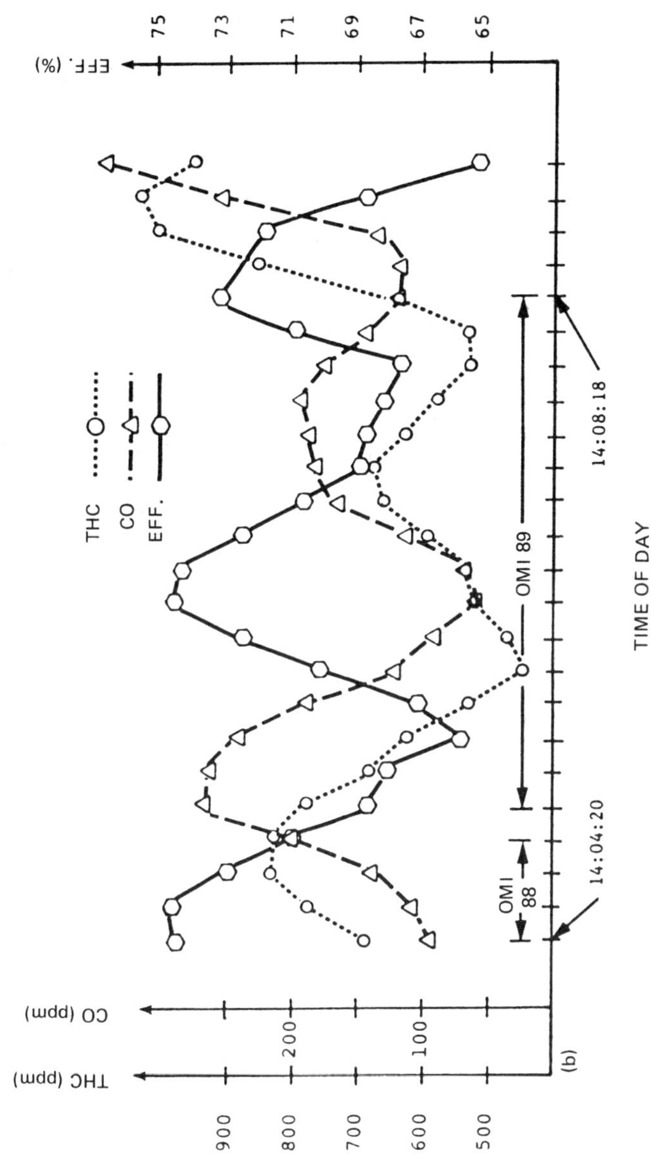

Fig. 10b Continuous probe sampling results, test 55.

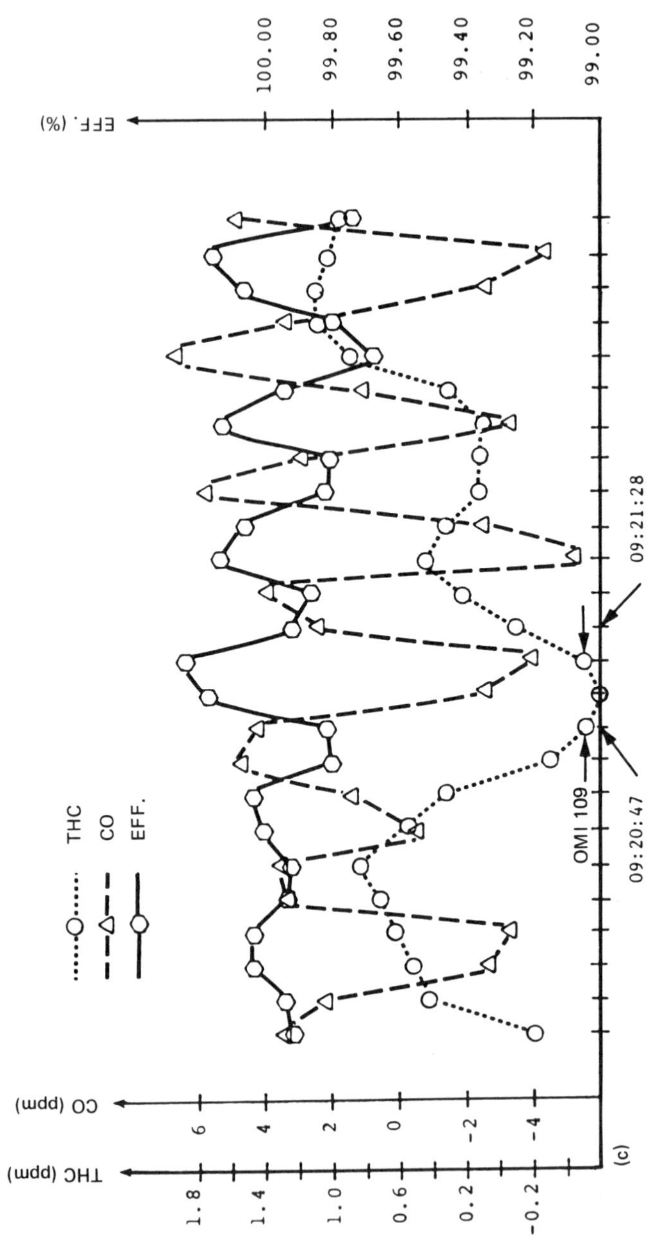

Fig. 10c Continuous probe sampling results, test 57.

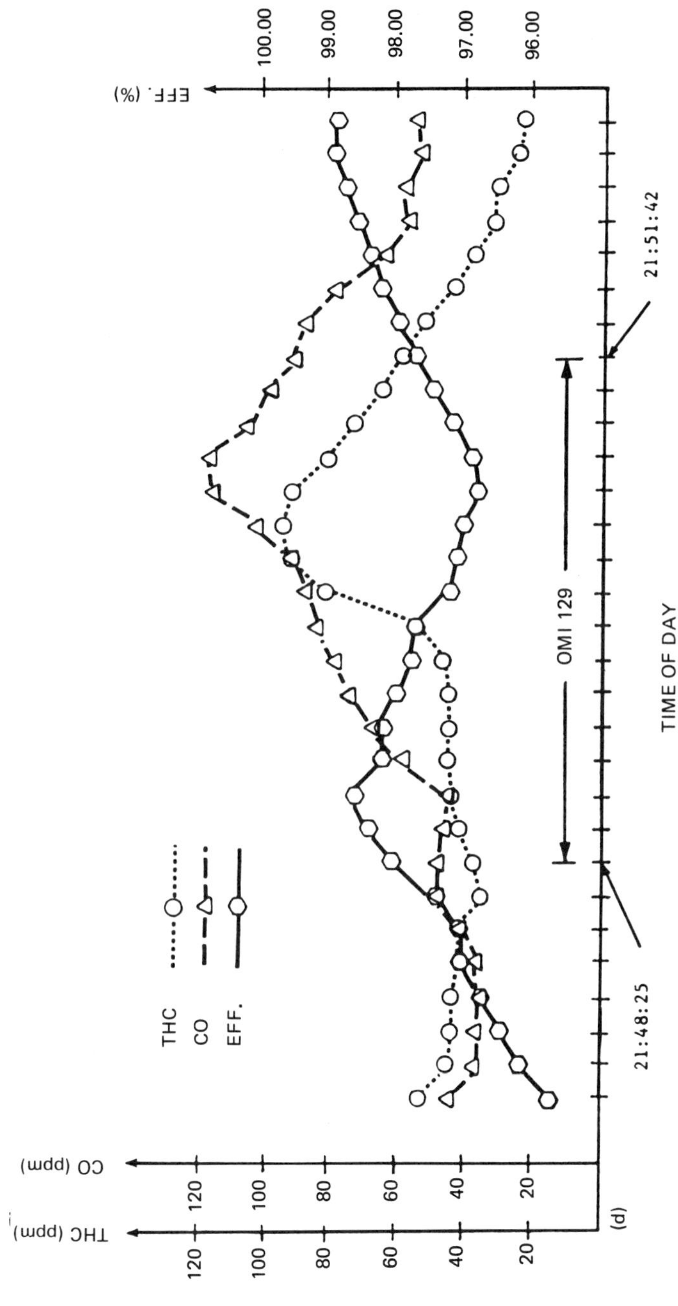

Fig. 10d Continuous probe sampling results, test 59.

sampling data for CO, CO_2, and THC. The efficiency shown is that calculated from Eq. (1) as

$$EFF \equiv \frac{[CO_2]}{[CO_2] + [CO] + [THC] + [Soot]} \tag{12}$$

The strong inverse correlation observed in this figure between efficiency and CO concentration demonstrates that the approximate efficiency formula of Eq. (2) is valid at high efficiency values. This appears to be the case because the total hydrocarbon concentration drops more rapidly than that of CO as efficiency increases. Consequently, at high efficiency, CO concentration becomes the driving parameter.

3.1 Concentration Calculations for Four Observed Field Tests

The procedure outlined above was applied to the four flare tests listed in Table 3. The selection of these specific samples was based on

(a) The observed signal-to-noise ratio of the data
(b) The desire for representative examples of both high and low efficiency cases, and
(c) The desire to include sample areas both coincident with the probe and within the flame itself

Figures 11 through 15 show the flare emission spectra which provided the data for the calculations. Clearly observed in all spectra is the CO_2 band appearing as red and blue spikes near 2250 cm^{-1} and 2375 cm^{-1}, respectively. Carbon monoxide is seen only in the low efficiency test (test 55 of Fig. 13) and in test 59 of Fig. 14, for which data were taken within the combustion zone itself. Superimposed on all the data are the best-fit, computer-generated spectra derived from using the procedures discussed above.

The discrepancies seen between observed and calculated data in the 2200–2250-cm^{-1} region of the red spike are attributable to deficiencies in the data base used by the ATLES code. These data base deficiencies are expected to lead to calculated results which underpredict the wings of the band at higher temperatures ($T \gtrsim 900$ K), as seen in Figs. 11, 12, and 14. At lower temperatures, the calculations should agree well with the observations (Fig. 13).

TABLE 3 Flare Tests for Concentration Calculation

Test No.	OMI No.	Height (ft)	Time (sec)	Efficiency[a] (%)
55	89	26	200	69.7
59[b]	129		200	97.7
50	50	25	56	99.1
53	68	18	40	99.0

[a]Average combustion efficiency from extractive probe measurements over same time interval as FT-IR measurements (McDaniel, 1983).
[b]Sample area included flame.

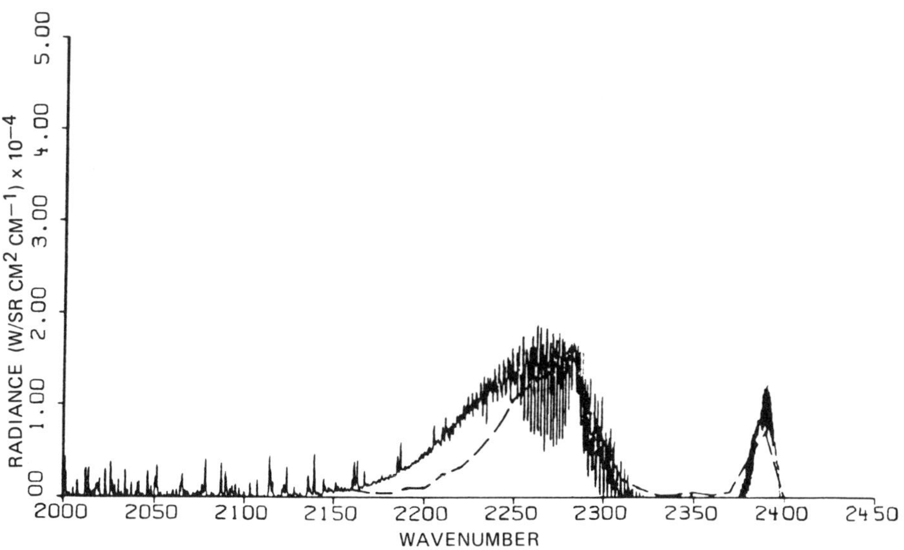

Fig. 11 Observed and calculated flare radiances for test 50. Deduced temperature is 920 K with 6000 ppm CO_2. Propylene flow rate is 159 lb/hr; steam flow rate, 0 lb/hr; nitrogen flow rate, 498 lb/hr.

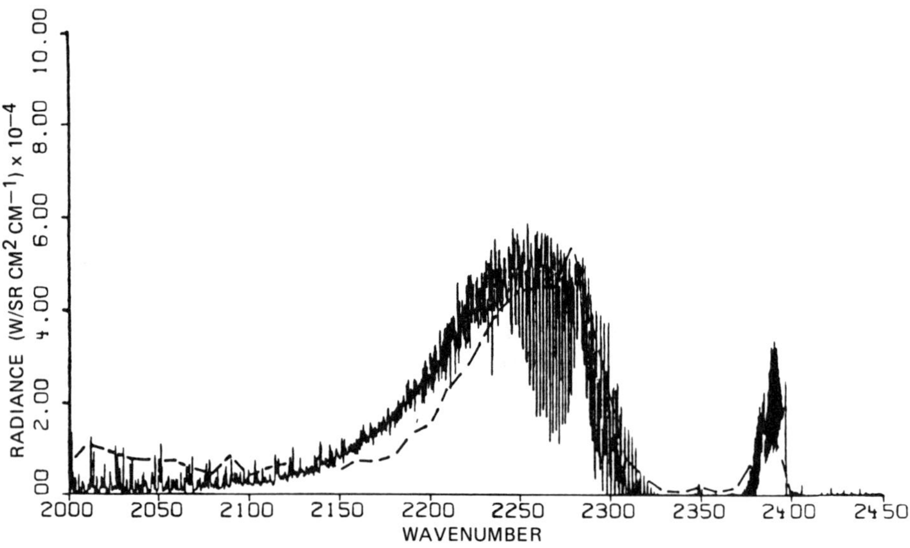

Fig. 12 Observed and calculated flare radiances for test 53. Deduced temperature is 1400 K with 2900 ppm CO_2 and 27 ppm CO. Propylene flow rate is 0.23 lb/hr; steam flow rate, 201 lb/hr; nitrogen flow rate, 1.46 lb/hr.

The plots of relative CO intensities used for initial temperature estimates in tests 55 and 59 are shown in Fig. 15. The final temperature and concentration values calculated for all four tests are shown in Table 4.

For comparison, the extractive probe results for these same tests are given in Table 5. Since the FT-IR integrates over a large volume while the probe extracts samples at a point, one would expect differences in the two results. Of the two data types, however, FT-IR gives the more spatially averaged result, since it represents a measurement through the total cross section of the plume, not just at a point within it.

3.2 Detection Limits

A number of factors in the measurement of flare efficiency by FT-IR have a potential impact on its accuracy. These include

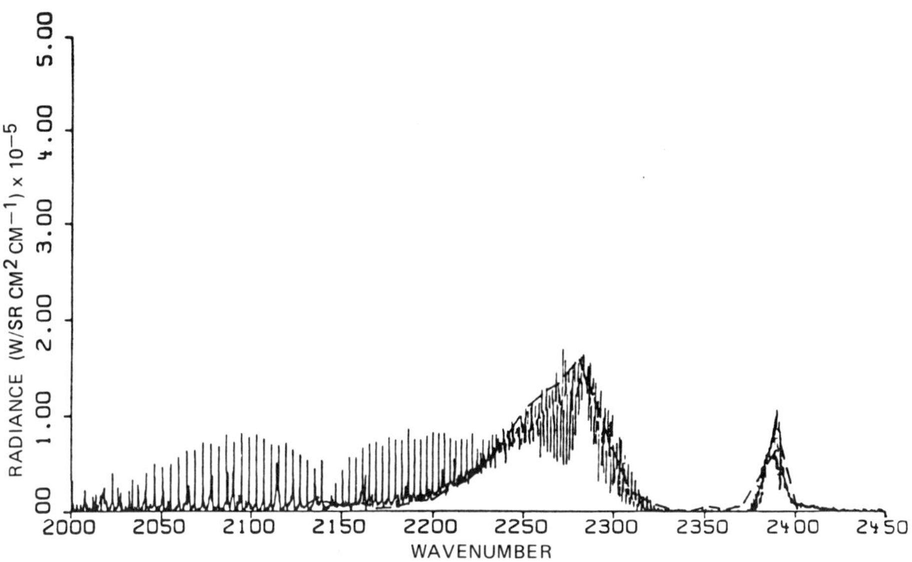

Fig. 13 Observed and calculated flare radiances for test 55. Deduced temperature is 683 K with 1800 ppm CO_2. Propylene flow rate is 161 lb/hr; steam flow rate, 0; nitrogen flow rate, 1125 lb/hr.

1. Temporal fluctuations in the flare
2. Spatial inhomogeneities within the field of view
3. Measurement noise inherent in the system
4. Accuracy of the computer models and their data bases used for analysis

Because of the limited data available, it is difficult to assess the contributions of these individual factors to the present FT-IR results. However, the comparisons of CO_2 concentrations made on tests 50 and 55 show a difference between probe and FT-IR results of only 15 to 30 percent. This can be considered very good agreement, since one would expect differences caused by the two distinct measurement techniques used, particularly considering the local nature of the probe measurement and the integrated nature of the FT-IR approach.

One quantitative indication of CO concentration detectability is obtained by noting that for a 20-ppm level, computer-predicted emission for CO just equals the noise in the measurement of test

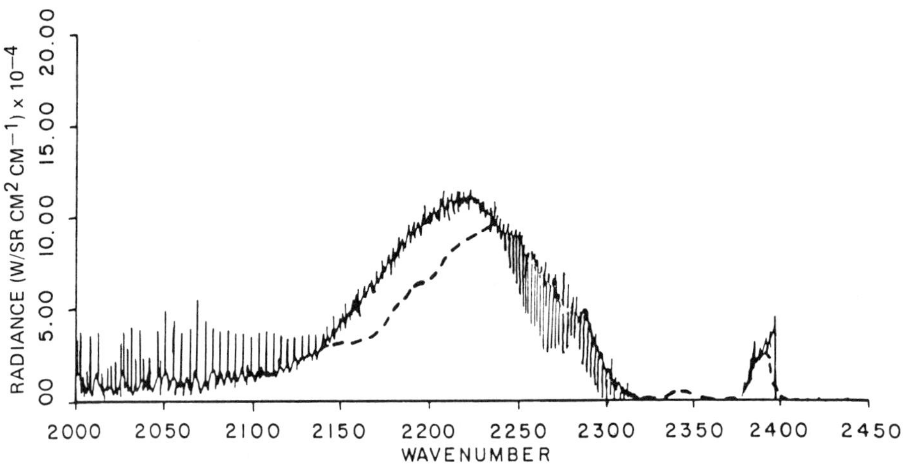

Fig. 14 Observed and calculated flare radiance for test 59. Deduced temperature is 2500 K with 6000 ppm CO_2. Propylene flow rate is 346 lb/hr; steam flow rate, 2133 lb/hr; nitrogen flow rate, 0 lb/hr.

50. (By comparison, in Chap. 1 the detection limit for the somewhat larger but much hotter flame was seen to be little less than 100 ppm, while for the considerably larger and cooler stack plume, it was less than 1 ppm.) Concentrations two to three times the 20-ppm level should therefore be reliably detectable. The detectability of methane was similarly calculated to be approximately 600 ppm, based on a noise equivalent radiance level of 1 W/cm^2 sr cm^{-1}. This exceeds even the 500-ppm level of methane resulting in the very low efficiency (70%) run of test 55. Note, however, that these values are dependent on the physical size (cross section) of the flare under test and would consequently be substantially lower for large industrial flares. Because the magnitude of the CO_2 concentration is sufficiently large, its minimum detectability is not of concern even for small flares.

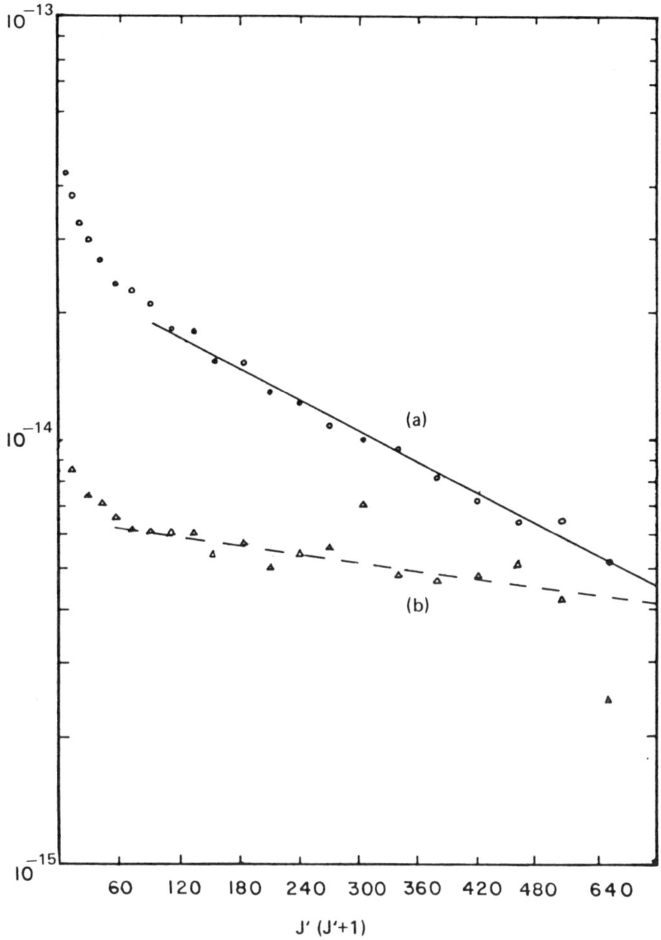

Fig. 15 Temperature estimates from CO intensities. Curve (a), test 55: slope = −3.04 E-3; T = 902 K. Curve (b), test 59: slope = 1.13 E-3; T = 2430 K.

TABLE 4 Flare Test Temperatures and Concentration (FT-IR)

Test No.	OMI No.	Temp. (K)	CO_2 (ppm)	CO (ppm)	EFF. (%)
55	89	683	1800	155	92.07
59[a]	129	2500	6000	1350	>81.63
50	50	920	>6000	<20	>99.67
53	68	1400	2900	~27	99.08

[a]FT-IR sample area 6 ft from flare tip within visible flame.

4 LABORATORY VALIDATION OF THE ATLES RADIATION PREDICTION CODE

4.1 Calculation of Plume Radiance

Evaluation of flare plume emission spectra and their propagation through atmospheric paths is complicated by the fact that the primary emitting species of the plume (H_2O and CO_2) are also present in the atmosphere. Their presence results in a high degree of correlation between the emitting and absorbing spectral lines, which must be taken into account when either calculating propagated radiance levels or when extrapolating observed data to conditions significantly different from those of the measurements. These correlation effects become particularly significant at moderate-to-low resolution, where the resultant radiance or transmittance is a mean value averaged over a finite spectral interval. Under these circumstances, the moderate resolution source radiance $\bar{N}_s$ multiplied by the moderate resolution atmospheric transmittance $\bar{\tau}_a$ is not equal to the true propagated radiance. That is,

$$\bar{N}_s \times \bar{\tau}_a \neq \overline{N_s \tau_a} \tag{13}$$

In general, the multiplicative result $\bar{N}_s \times \bar{\tau}_a$ will yield a propagated radiance level substantially larger than the true value $\overline{N_s \tau_a}$. To treat this problem properly (at moderate-to-low resolution), one must therefore treat the entire optical path (plume and atmosphere) as a single highly inhomogeneous path.

From a theoretical point of view, the problem of calculating plume radiance as propagated through an intervening atmospheric path can be treated by either high resolution, line-by-line (Randall, 1974) HITRAN methods or by inhomogeneous-path band model (Young, 1977a) methods. Unfortunately, both of these methods suffer from an inherent limitation in the data bases on which they depend. In the case of line-by-line calculations, the required data consist of individual spectral line positions, widths, strengths, and lower state energies. The most comprehensive set of such data is the AFGL atmospheric absorption line compilation. As the name implies, however, the AFGL compilation was formulated primarily for atmospheric absorption problems and lacks the excited state bands and/or higher rotational lines required for application at temperatures in excess of 500–700 K. For band model calculations, the required data base consists of tabulated values of the mean line density and the mean line-strength-to-line-spacing ratio as a function of temperature and frequency. Unlike the line-by-line parameters, band model parameters are available (Ludwig et al., 1973) for temperatures in excess of 1200 K*; also they are calculable from the AFGL compilation for temperatures below approximately 500 K. For the intermediate temperature regime of significance to many flare tests, however, the only set of parameters available is that extrapolated from the two basic sets.

As one would expect, the lack of intermediate temperature parameters is accompanied by a lack of controlled intermediate temperature data for either validating the models or for determining new parameters. In an attempt to generate a comprehensive data set that can be used for both radiance level determinations and validation of current band modeling methods, 4-cm^{-1} resolution spectra were generated from 1150 cm^{-1} to 2550 cm^{-1} at temperatures of 500, 700, and 900 K. The data consist of source radiance, atmospheric transmittance, and propagated radiance spectra, at each of these temperatures, simulating altitudes of 0, 2, and 5 km, using atmospheric path lengths of 252, 504, and 1008 m.

---

*The elevated parameters available for CO_2 are based on calculations; those for H_2O have been determined by direct emission–absorption measurements.

TABLE 5 Extractive Probe Data

A. Continuous samples				
Test No.	50	53	55	59
OMI No.	50	68	89	129
Time—begin	14:09:36	16:25:45	14:05:13	21:48:25
Time—end	14:10:28	16:26:26	14:08:18	21:51:42
Temp. (K)	415	387	343	421
CO (ppm)	35.04	17.9	124.6	79.9
CO_2 (ppm)	5382	2438	1371	5965
THC (ppm)	16.8	7.0	475.3	62.9
O_2 (ppm)	20.10	19.38	20.85	19.26
SO_2 (ppm)	0.040	0.648	0.085	0.039
NO_x (ppm)	0.60	1.10	0.13	1.30
EFF. (%)	99.06	98.98	69.68	97.69
Ambient temp. (K)	301	303	304	295
Wind (mph)	1.8	2.6	2.3	0.8
Bkg. CO (ppm)	0.1	0.3	0.2	0.0
Bkg. CO_2 (ppm)	415	375	515	421
Bkg. THC (ppm)	5.4	5.7	6.8	2.9
Bkg. O_2 (ppm)	20.81	20.50	21.23	20.61
Bkg. SO_2 (ppm)	0.009	0.050	0.016	0.002
Bkg. NO_x (ppm)	0.12	0.14	0.78	0.10
B. Instantaneous samples				
Test No.	50	53	55	59
OMI No.	50	68	89	129
Methane (ppm)	4.45	3.38	14.07	4.09
Ethane/ethylene (ppm)	1.77	0.63	3.35	1.89
Acetylene (ppm)	0.88	0.25	0.20	0.51

Table 5 (continued)

Propane (ppm)	0.76	2.23	268	3.97
Propylene (ppm)	3.29	9.13	1075	1.63
Time	14:11:08	16:26:49	14:16:29	21:48:18
	C. Integrated bag samples			
Test No.	50	53	55	59
OMI No.	50	68	89	129
Methane (ppm)	4.27	4.05	16.9	4.89
Ethane/ethylene (ppm)	3.57	4.05	40.8	9.10
Acetylene (ppm)	1.22	0.53	23.2	2.85
Propane (ppm)	4.94	3.99	277	13.4
Propylene (ppm)	20.7	17.0	1141	57.5
Time	14:03- 14:11	16:25- 16:38	11:09- 11:16	21:47- 21:54

Source: McDaniel (1983).

4.2 Band Models as Predictors

At high temperatures, band models that are formulated from a statistical representation of the spectral lines become more appropriate because of the larger number of significant spectral lines per unit frequency interval. In addition, the parameters required for the band models may be evaluated in a relatively straightforward manner, regardless of the complexity of the spectrum. With the current state of the art, band models are the only viable approach to prediction of elevated temperature gaseous emissions, especially at temperatures in excess of about 700 K, at which the high resolution parameters begin to fail.

Although it would appear that band models are ideally suited to the plume propagation problem, some difficulties are associated with their use. The most obvious one is that these models are inherently limited in spectral resolution. Because they operate at finite resolution, the correlation between emitting and absorbing spectral lines, discussed above, makes the overall radiative transfer problem inseparable. The total path then (consisting of the atmosphere and elevated temperature source) must be treated as a single, highly inhomogeneous one with the equations of transfer being integrated over this total path.

The basic equation defining radiative transfer for a general inhomogeneous path is

$$\bar{N}_{\Delta\bar{\nu}} = -\int_0^l N^0(\bar{\nu}_0, s) \frac{d\bar{\tau}(\bar{\nu}, s)}{ds} ds \tag{14}$$

where $\bar{N}_{\Delta\bar{\nu}}$ is the mean radiance in a spectral interval $\Delta\bar{\nu}$ about $\bar{\nu}_0$ for the total path beginning at coordinate 0 and ending at coordinate l. In the integral $N^0(\bar{\nu}_0, s)$ is the Planck function, which is assumed to be reasonably constant in the interval $\Delta\bar{\nu}$; $\bar{\tau}(\bar{\nu}, s)$ is the mean transmittance in the spectral interval and is evaluated for the path 0 to s. For a band of spectral lines that may be treated as randomly positioned lines of equal width, the statistical band model is appropriate. For this model it may be shown (see Chap. 5, the discussion preceding Eq. (27), and also Goody, 1964) that

$$\bar{\tau}(\bar{\nu}, s) = \exp[-\bar{W}(s)/d] \tag{15}$$

where $\bar{W}(s)$ is the mean equivalent width of the spectral lines in the spectral interval and d the mean spacing of these lines. With this definition, the transmittance derivative in Eq. (14) may be evaluated, using

$$\frac{d\bar{\tau}(\bar{\nu}, s)}{ds} = -\bar{\tau}(\bar{\nu}, s) \frac{1}{d} \frac{d\bar{W}(s)}{ds} \tag{16}$$

The problem of evaluating Eq. (14) therefore reduces to that of evaluating the mean equivalent width derivative $d\bar{W}(s)/ds$; once this derivative is obtained, it may be integrated along the path to give $\bar{W}(s)$ and thus $\bar{\tau}(\bar{\nu}, s)$. For an isolated spectral line, the equivalent width W is the integrated area under the absorption profile, or

$$W(s) = \int_{-\infty}^{\infty} \left\{ 1 - \exp\left[-\int_0^s \chi(s') p(s') \kappa_0(\bar{\nu}, s')\, ds' \right] \right\} d\bar{\nu} \tag{17}$$

where $\chi(s')$ is the concentration (as mole fraction) of the species of interest at path position s', p(s') the total pressure of this point, and $\kappa_0(\bar{\nu}, s')$ the absorptivity (per column density of gas in atm·m) for the line shape assumed. Because Eq. (17) is the value for an isolated spectral line, it is obviously not the mean value required for Eq. (15). However, for equal width lines, the equivalent width is strictly a function of the line strengths. By postulating a distribution function for the line strengths, one may therefore determine the mean equivalent width, using

$$\bar{W}(s) = \int_0^\infty \mathscr{P}(S) W(s, S)\, dS \tag{18}$$

where $\mathscr{P}(S)$ is the probability that a random line in $\Delta\nu$ has strength between S and $S + dS$.

Band model formulations therefore eliminate the need for explicit knowledge of (a) individual linewidths by assuming an array (or arrays) of equal width lines and (b) individual line strengths by postulating a statistical distribution of strengths in the band. Even with these simplifications however, the task of evaluating Eq. (17) is tedious, since an integration over the specific path of interest is required as well as an integration over frequency. Because it would be desirable to perform the frequency integration once and for all and to reevaluate $W(s)$ only as required for specific paths, band model formulations have attempted to separate these integrals effectively, allowing for a reversal of the orders of integration.

4.3 Band Model Calculations

Prior to the flare tests discussed in this chapter, observed experimental data were compared with the statistical models to validate the ATLES code (Young, 1977) as modified to operate on an HP-1000 computer system. This code allows for the use of Lorentz, Doppler, or Voigt line shapes as well as selection of Curtis–Godson or derivative (Lindquist–Simmons, 1972) models. For all calculations presented here, the Lorentz line profile was selected because the pressures encountered were high enough to ensure dominance of pressure broadening. For all atmospheric path transmittance or source radiance calculations for which there is a high degree of path inhomogeneity, the Lindquist–Simmons derivative model was employed (Young, 1977).

4.4 Parameters for Band Models

The input data required for all of these models are a tabulation of the frequency- and temperature-dependent parameters $\bar{K}$ and β or $\bar{K}$, $\bar{\gamma}$, and d, depending on which model is being used, where $\bar{K} = \overline{S/d}$ the mean line strength to line spacing ratio, $\bar{\gamma}$ the mean linewidth in the resolution element of interest, d is the mean line spacing in this element, and β the standard band model parameter defined as $2\pi\bar{\gamma}/d$. At the present time, four major sets of parameters are available for the 4.3-μm CO_2 and 6.3-μm H_2O bands. These are

1. The NASA/General Dynamics parameters primarily applicable to elevated temperatures ($T \gtrsim 1200$ K)
2. Parameters derived from the AFGL atmospheric absorption line compilation (McClatchey et al., 1973; see Appendix for availability) generally useful only at temperatures below about 700 K
3. Parameters derived from molecular theory and hence valid for various temperature ranges
4. Hybrid parameter sets derived by extrapolation of 1, 2, and/or 3 to form sets with broader temperature coverage

The NASA/GD parameters for H_2O, derived from emission-absorption measurements using a long strip burner, are primarily applicable only at temperatures in excess of 1200 K. All lower temperature H_2O parameters in this set, derived from extrapolation of the high temperature data (as well as from analyses of low temperature spectra), are less applicable. Parameters for the 6.3-μm band are given at 25 cm^{-1} resolution covering the range from 300 to 3000 K for $\bar{K}$ and covering from 600 to 3000 K for d. Young (1974) has used a semilogarithmic linear extrapolation to expand the line spacing data below 600 K.

For CO_2, the NASA/GD parameters are the result of theoretical calculations using the known spectroscopic constants for carbon dioxide to evaluate line positions. Calculations also used harmonic oscillator approximations to determine hot band strengths from experimental ground state values. In the 4.3-μm region, both $\bar{K}$ and d values are available throughout the 300–3000 K range at 5- and 10-cm^{-1} resolution.

Since the AFGL absorption line compilation tabulates line positions, strengths, half-widths, and lower state energies, band model parameters may be derived from this compilation by appropriately averaging over the desired spectral intervals. Using the lower state energies given, the line strengths may also be corrected to cover any desired temperature range. Since the compilation was assembled to treat atmospheric problems, however, it does not contain the hot-band lines or high rotational state lines which become significant at elevated temperatures. The lack of these lines and bands places an upper limit on the range of validity for parameters derived from the compilation. Numerous comparisons of both line-by-line and band model calculations with experimental data indicate that the upper temperature limit for this compilation is on the order of 600–700 K.

From the preceding discussion, one can see that the NASA/GD parameters and the AFGL-generated parameters are highly complementary: the higher temperature regime is covered by the NASA/GD set and the lower temperature regime is covered by the AFGL-generated set. Because they are complementary, sets of parameters have been formed by extrapolating and interpolating these two types of parameters to yield hybrid sets that are more appropriate for the entire temperature range from 300 to 3000 K.

The parameters used for the 4.3-μm carbon dioxide band are a hybrid set derived by S. J. Young (1976) that combine the NASA/GD parameters with those derived from the AFGL tabulation. This hybrid set is included as part of the data base in the ATLES code and is called the COMBCO2A parameter file. The file contains tabulated values from 2000 cm^{-1} to 2400 cm^{-1} at 5 cm^{-1} resolution and spans the temperature range from 100 to 3000 K. A new set of band model parameters for CO_2 have recently become available (Bernstein, 1980) which remedy some of the deficiencies found in the COMBCO2A set. This new set is incorporated in the SIRRM (Standardized Infrared Radiation Model) code (Ludwig et al., 1981) a generally useful source of band model information.

Because all of the current measurements (500, 700, and 900 K) were taken at temperatures below the primary range of validity for the NASA/GD H_2O parameters and because there was no comparable hybrid set available for the 6.3-μm band of this molecule, calculations performed in this region used the theoretically derived parameters of L. S. Bernstein. These were derived using two basic assumptions for A-type or B-type bands: (a) that the shapes of similar bands for water vapor may be superimposed and (b) that the integrated band intensities for bands in a given sequence may be related by harmonic oscillator matrix elements. With these assumptions, the line positions for lines in an unknown band could be computed directly from those of a well-known reference band by simply shifting the lines by the band center displacements; the integrated line strengths could be determined from the ratio of the known-to-unknown harmonic oscillator matrix elements and population factors.

The parameter set supplied to the authors by Bernstein was derived in the manner outlined above, using as reference water bands the 000 → 010 and 010 → 100 B-type bands and the 010 → 001 A-type band. In all cases, the positions and strengths for these reference bands were taken from the updated version of the AFGL atmospheric absorption line compilation containing recent additions to the 6.3-μm H_2O transitions as documented by Flaud and Camy-Peyret (1976) and Flaud, Camy-Peyret, and Mandin (1977). The parameter set contained $\bar{K}$ and d parameters for H_2O covering the 1010–2500-cm^{-1} region at temperatures from 275 K to 2400 K at a spectral resolution of 10 cm^{-1}. The additional parameter $\bar{\gamma}$ required by the ATLES code (the mean broadening parameter for nonresonant self-broadening at STP) was derived by averaging H_2O linewidths from the same AFGL tabulation; these values were then converted from foreign-broadened values to nonresonant self-broadened values, assuming that γ_{self} = 1.170 γ_{air}.

4.5 Comparisons with Experimental Data

Using the ATLES code and the parameter sets discussed above, calculations of source radiance, simulated atmospheric path transmittance, and propagated radiance were performed to compare to the emission spectra (signatures) and propagation measured under the conditions outlined in Table 6. For the simulated atmospheric path, which was isothermal and isobaric, a single homogeneous segment was used in the model. For the source or hot-cell path, which was nonisothermal, the temperature profiles of Fig. 16 were used initially with the total one-meter length subdivided into twenty-one segments. Preliminary calculations of source radiance based on these profiles, however, were found to seriously overpredict the observed values at both 700 and 900 K. The profiles shown in Fig. 16, however, were those measured within the furnace used to heat the cell containing the emitting gas but along the exterior wall of the cell. We found that these temperatures were somewhat higher and exhibited a somewhat milder gradient for the 700 and 900 K profiles than those observed with three thermocouples measuring the actual interior gas temperature at the center and ends of the cell. As a result, profiles were generated which passed through the observed

TABLE 6 Fill Conditions for Aircraft Signature Propagation Measurements

	Sea level		2-km altitude		5-km altitude	
	Hot cell	White cell	Hot cell	White cell	Hot cell	White cell
Temperature (K)	500,700,900	296	500,700,900	296	500,700,900	296
Water vapor (Torr)	41.04	14.0	33.09	5.82	22.86	0.925
Carbon dioxide (Torr)	25.08	0.25	19.86	0.199	13.72	0.137
Total pressure	760	760	601.7	601.7	415.6	415.6

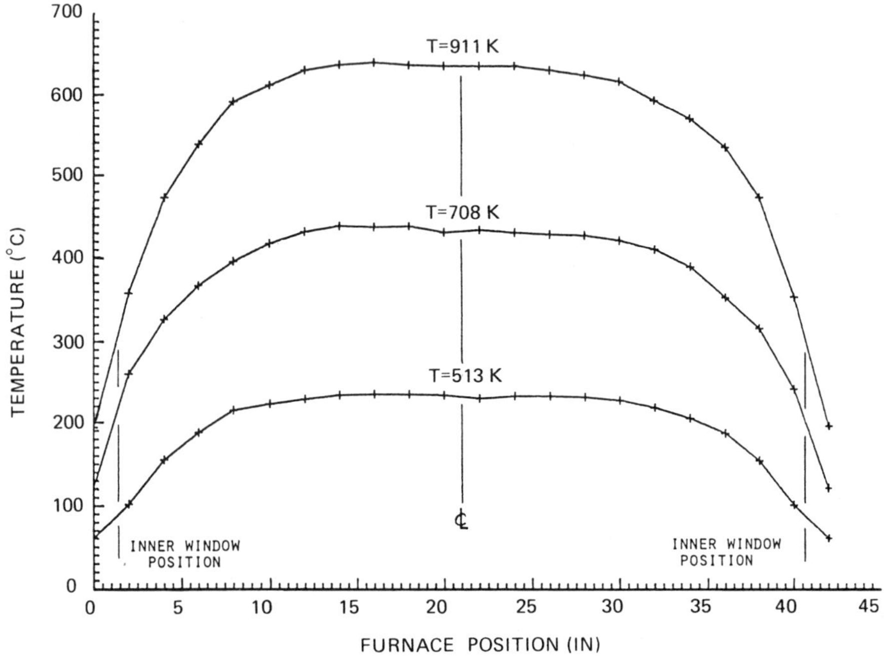

Fig. 16 Temperature profiles for hot cell at 513, 708, and 911 K.

thermocouple temperatures and matched as closely as possible the shapes of the original profiles. The results obtained with both the original and modified profiles for nominal 700 and 900 K peak temperatures* as compared to the observed data for the three simulation altitudes measured (0, 2, and 5 km) are shown in Figs. 17, 18, and 19.

As may be seen from these figures, the modified profiles had a far more pronounced effect on the 700 K spectra than on the 900 K spectra. In general, at all three simulation altitudes the H_2O (1150–2150 cm^{-1}) and CO_2 (2200–2400 cm^{-1}) calculations for the 700 K source agree very well with the observed data. Some slight overprediction remained, however, in the water bands (increasing somewhat with simulation altitude), and some underpre-

*The original 500 K profile agreed with the thermocouple reading so that no alteration was required in this case.

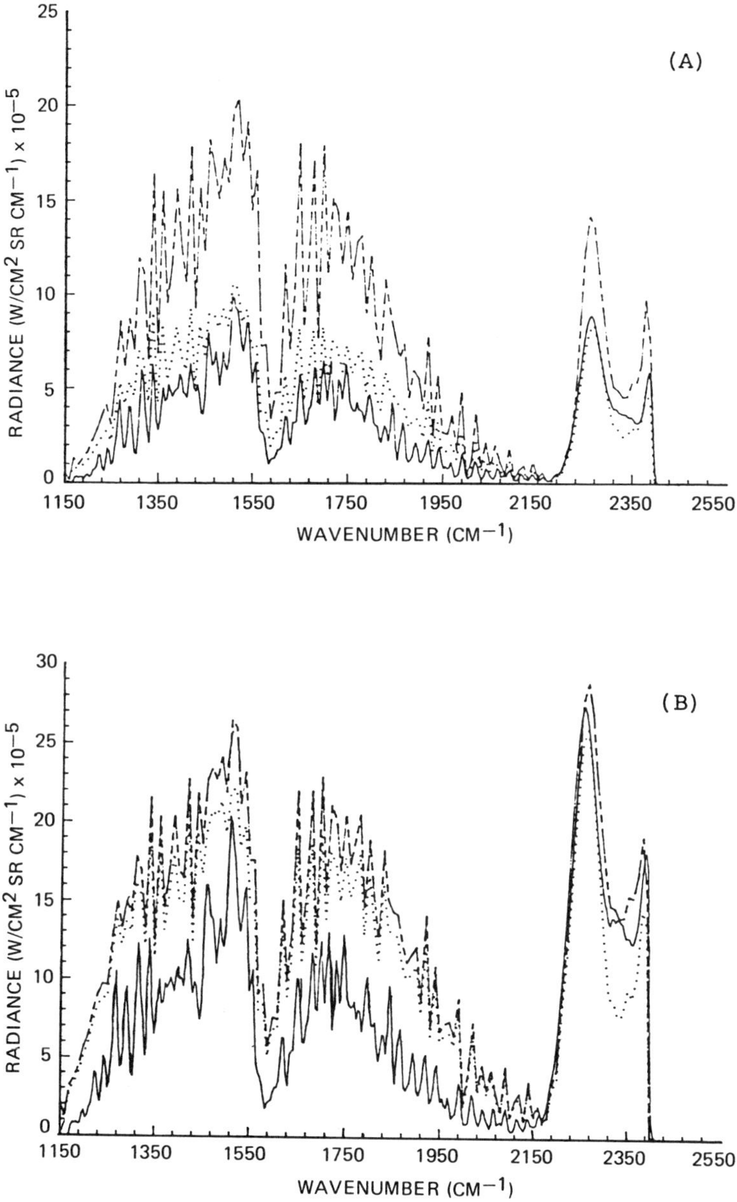

Fig. 17 Observed source radiance (—) compared to calculated values using the original (− • —) and modified (• • •) temperature profiles. Simulated altitude is 0 km, source temperatures are 700 K (A) and 900 K (B).

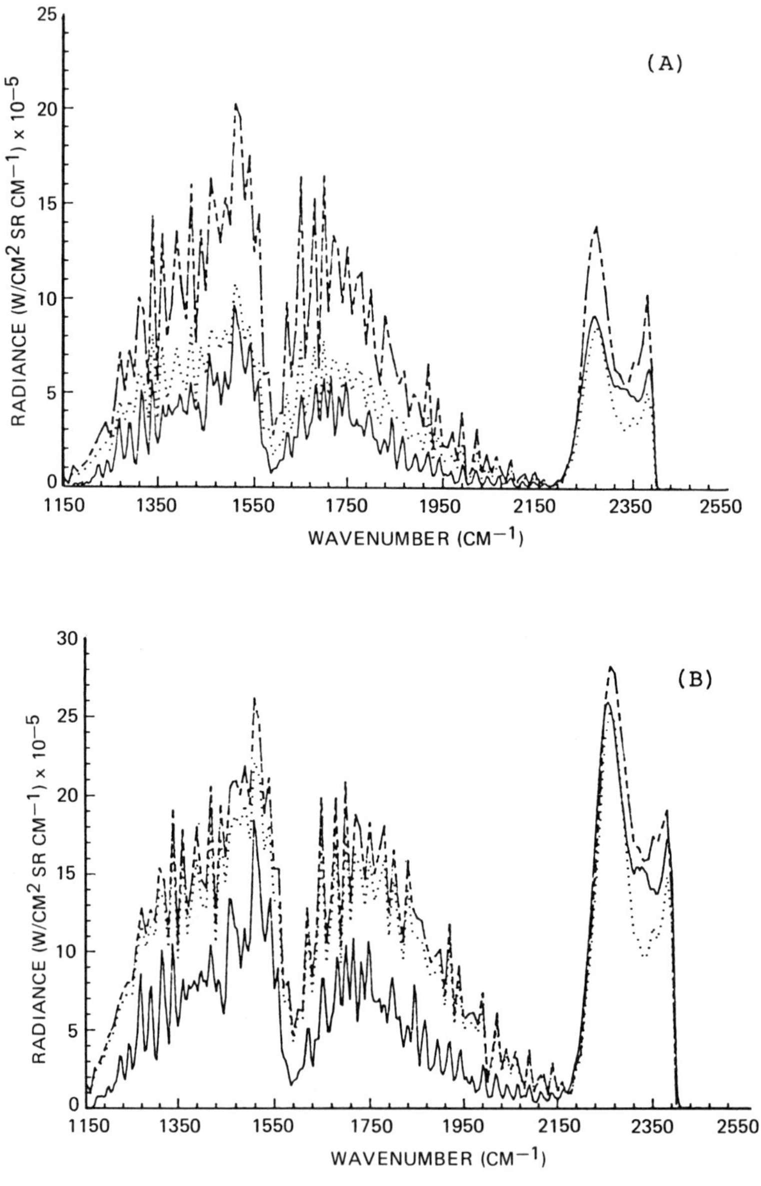

Fig. 18 Observed source radiance (—) compared to calculated values using the original (−•−) and modified (•••) temperature profiles. Simulated altitude is 2 km, source temperatures are 700 K (A) and 900 K (B).

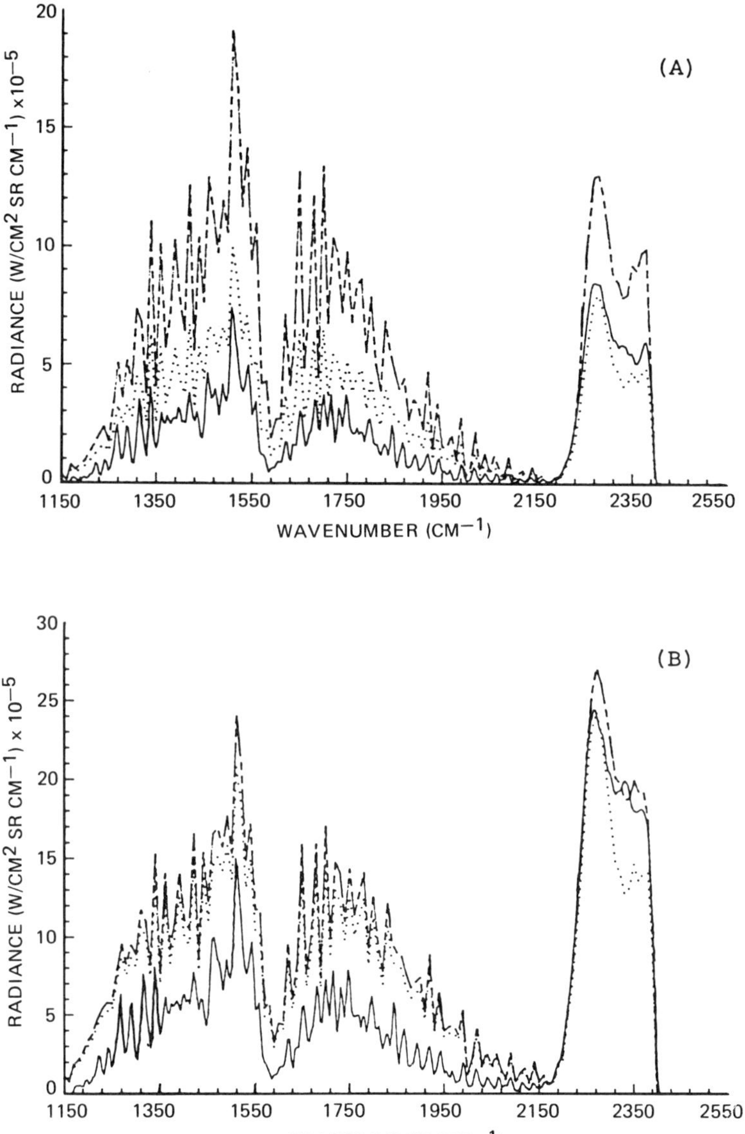

Fig. 19 Observed source radiance (—) compared to calculated values using the original (– • –) and modified (• • •) temperature profiles. Simulated altitude is 5 km, source temperatures are 700 K (A) and 900 K (B).

diction occurred for the CO_2 bands (remaining relatively invariant with altitude). For the 900 K source temperature, however, the calculated CO_2 emissions went from slightly overpredicting the observed values to slightly underpredicting them; the water vapor spectra showed only small downward displacements remaining substantially above the measured curves. In contrast to this behavior, the reversals observed in the CO_2 bands near 2330 cm^{-1} (caused by self-absorption of the CO_2 emissions by the cooler periphery gases) are augmented substantially by the modified 900 K profile but are changed little, if any, at 700 K. This effect, probably caused by the more rapid gradient in the 900 K case, gives rise to a larger decreased temperature region and, hence, to more severe line reversal.

Calculations of the atmospheric path transmission agreed very well with the observed spectra for the CO_2 bands and exhibited a somewhat larger transmission (lower line density or line strength) than required in the water vapor bands. Representative examples of the calculated and observed transmission curves are shown in Figs. 20, 21, and 22 for path lengths of 252 and 504 m at altitudes of 0, 2, and 5 km. Because both the CO_2 and water vapor parameters at low temperatures were derived from the AFGL compilation, one would anticipate relatively good agreement for the transmittances. The discrepancies in the 6.3-μm water bands are somewhat larger than expected, but these could arise from several sources such as a lack of significant spectral lines in the AFGL tabulation or a breakdown of the equal linewidth or band shape assumptions used in deriving the parameters.

The propagated radiances obtained from the ATLES code, examples of which are shown in Figs. 23 through 25, showed a consistent overprediction for the water vapor bands and a slight underprediction in the CO_2 bands. The overpredictions observed in the water vapor bands, which tend to increase with increasing temperature, are probably indicative of deviations from the harmonic oscillator dependence assumed for the bands of a given sequence in deriving the parameters. The fact that the calculations are within a factor of two to three for the observed values under all conditions observed does imply that the procedures used by Bernstein, with some modification, should be capable of yielding a reasonably accurate set of parameters for this band. The deviations observed in the CO_2 bands, on the order of only 20% to 30%, are not highly significant from the point of view of the band model itself. The extreme sensitivity of the propagated radiances in this band, to both the source temperature and to the particular gradients present, could easily account for deviations on this

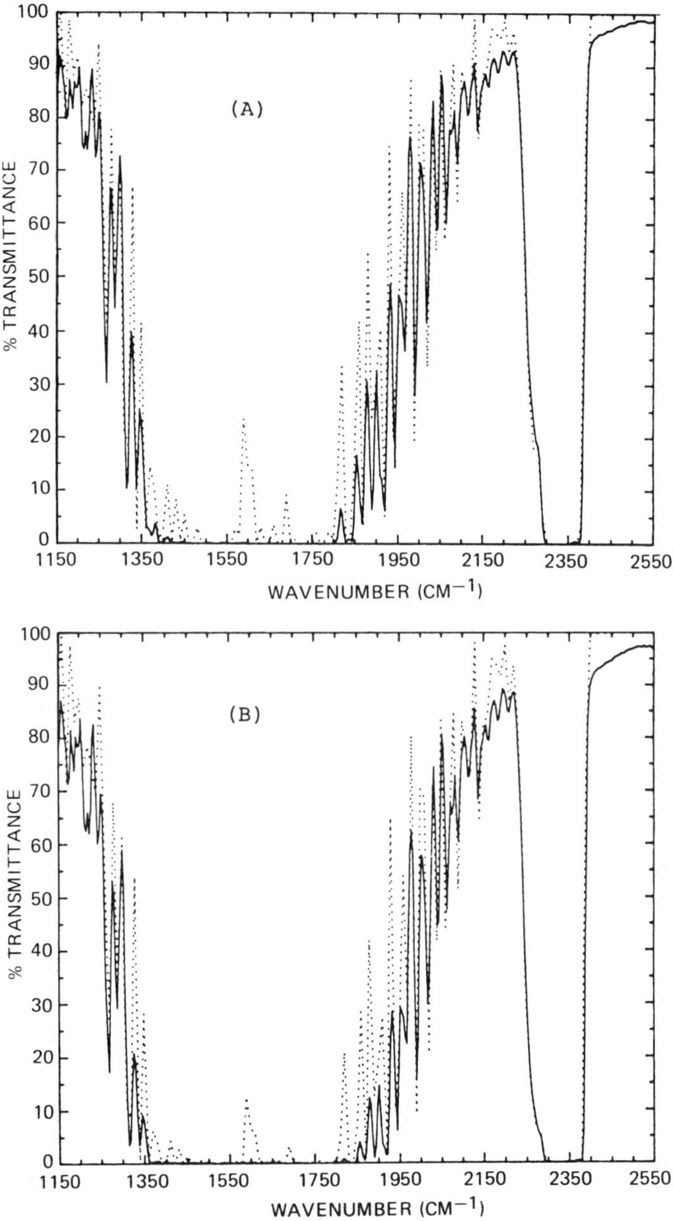

Fig. 20 Observed and calculated atmospheric path transmittance at an altitude of 0 km over path lengths of 252 m (A) and 504 m (B).

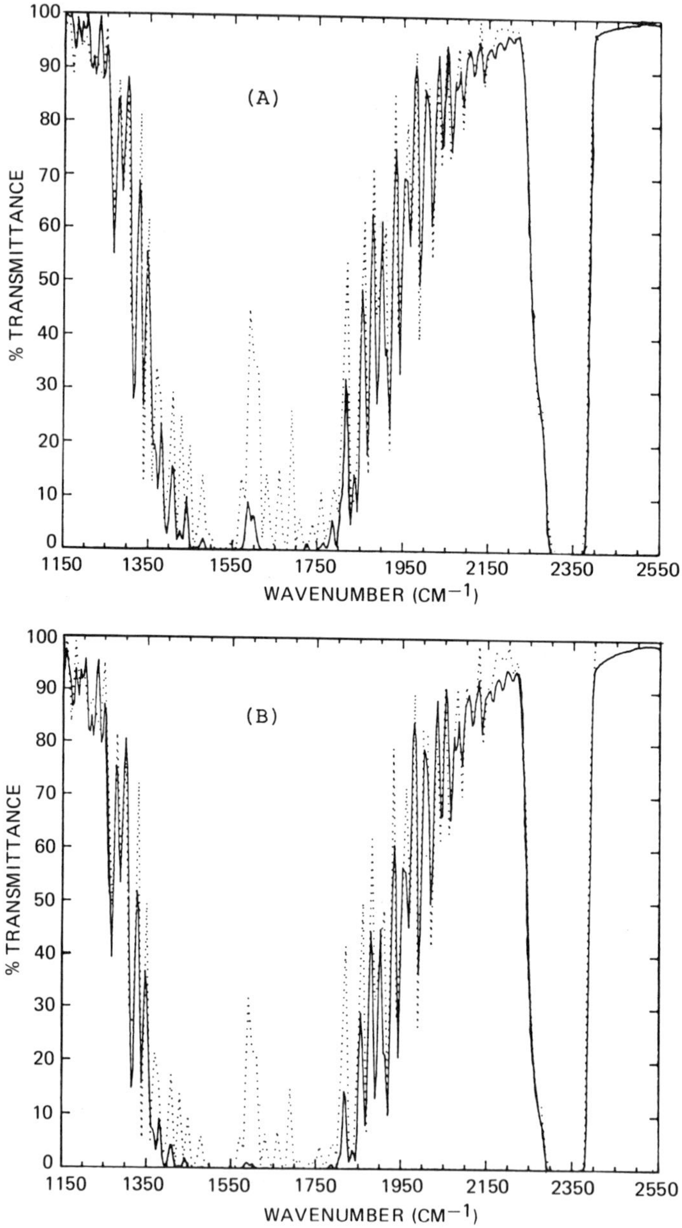

Fig. 21 Observed and calculated atmospheric path transmittance at an altitude of 2 km over path lengths of 252 m (A) and 504 m (B).

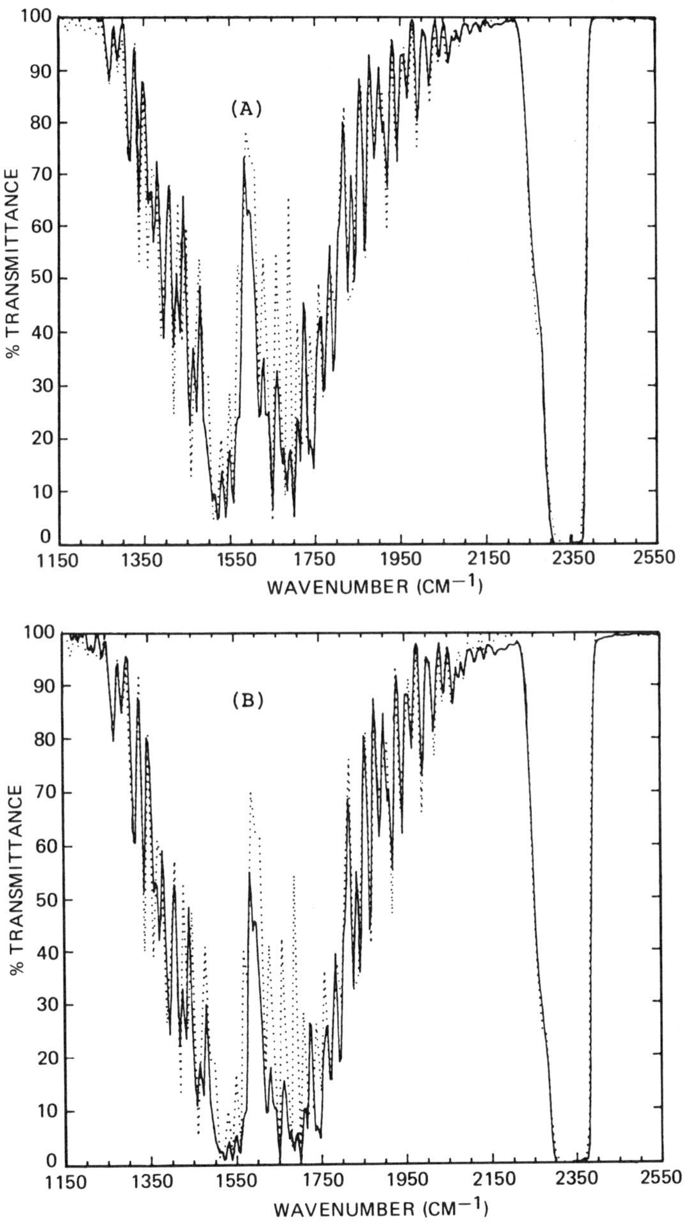

Fig. 22 Observed and calculated atmospheric path transmittance at an altitude of 5 km over path lengths of 252 m (A) and 504 m (B).

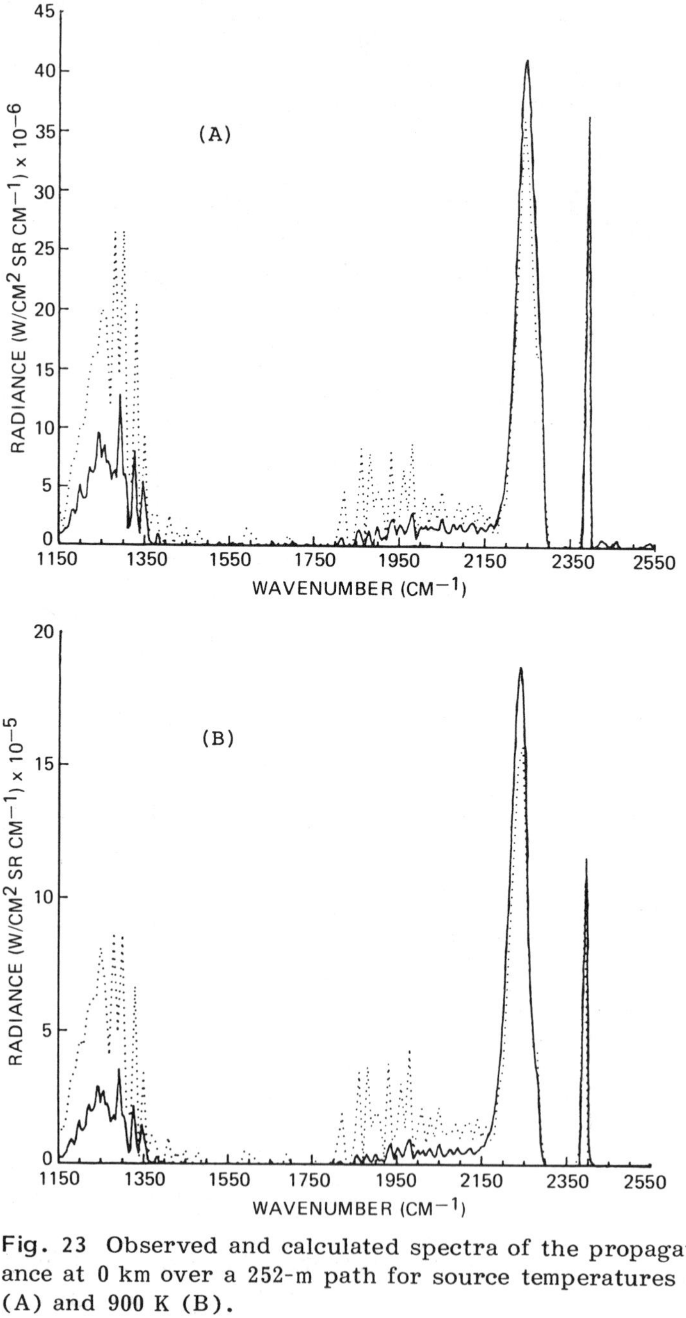

Fig. 23 Observed and calculated spectra of the propagated radiance at 0 km over a 252-m path for source temperatures of 700 K (A) and 900 K (B).

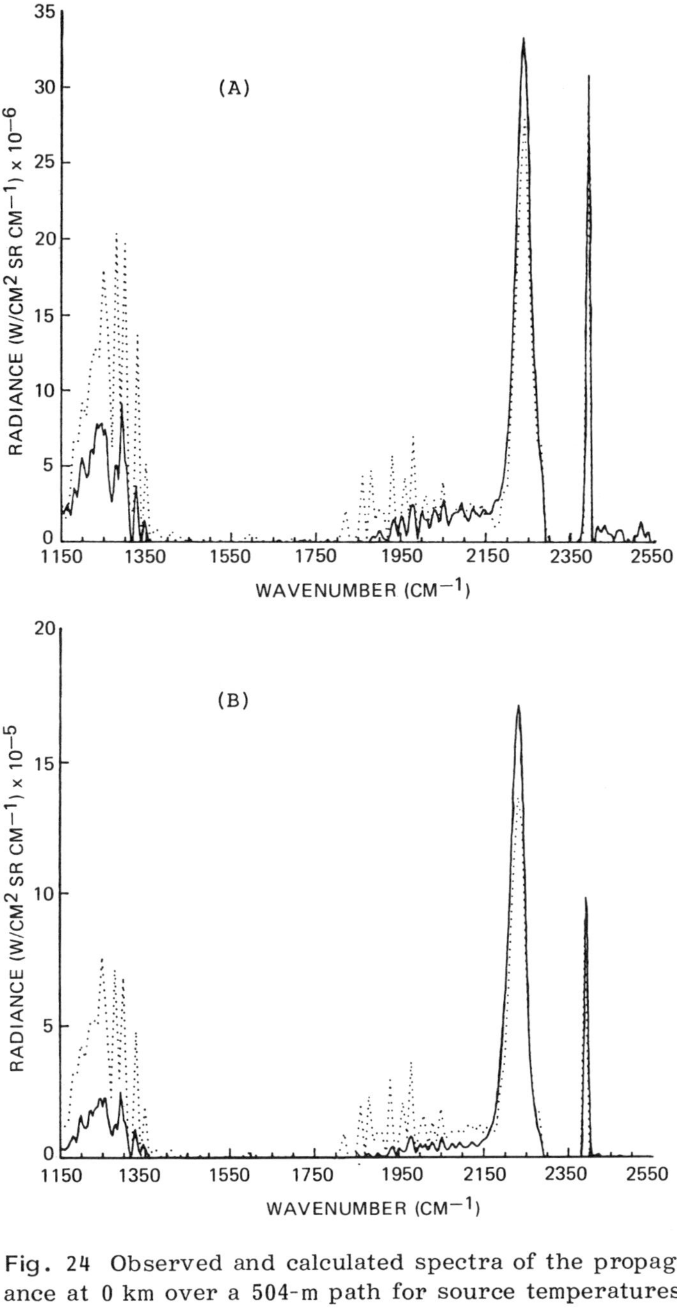

Fig. 24 Observed and calculated spectra of the propagated radiance at 0 km over a 504-m path for source temperatures of 700 K (A) and 900 K (B).

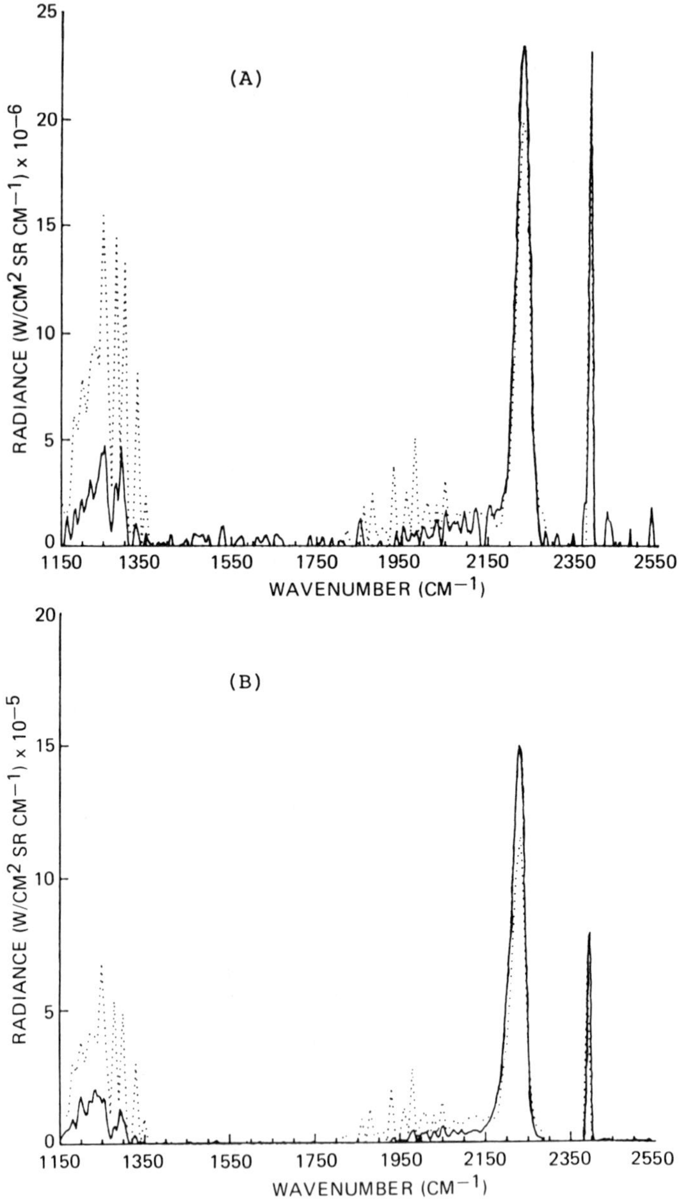

Fig. 25 Observed and calculated spectra of the propagated radiance at 0 km over a 1008-m path for source temperatures of 700 K (A) and 900 K (B).

order. In addition, the parameters in the data set used for these calculations are the result of extrapolating the NASA/GD- and AFGL-generated values into the intermediate temperature range to form a single contiguous set. Extrapolations are required throughout the 600–1000 K range and are most severe in the wings of the bands (Young, 1976). It is highly probable that the parameters themselves are somewhat in error. This may be particularly true in the wings of the bands which, unfortunately, contribute most heavily to the propagated radiance spikes.

5 CONCLUSION

Because the ATLES radiation code is shown to be basically valid, the FT-IR method is a realistic approach to combustion efficiency monitoring. In general, the results of the comparison between point sampling of combustion products and remote sensing of those products using IR spectroscopic techniques clearly indicate that IR methods can offer substantial advantages over currently available methods. Even with the limited developmental work possible, baseline procedures were formulated allowing for evaluation of plume temperature, CO concentration, and CO_2 concentration based solely on the IR spectra. With these values, estimates of efficiency could then be made, using the approximate relationships of Eq. (2), which at high efficiency gave values in good agreement with the point sampled results (Table 7). The only major shortcoming of the IR approach was its inability to determine total hydrocarbon concentrations necessary for monitoring lower efficiency flares. This was the result of using a general IR instrument not specifically tailored to this measurement problem as well as the result of looking at a small flare, with limited optical depth, rendering difficult the detection of weaker IR emitters. Given a solution to this problem, however, IR methods look attractive because they allow for remote, nonhazardous, nonintrusive monitoring of industrial type flares at distances of 1/4 to 1/2 mile. The methodologies developed also look straightforward enough that they could be implemented on field hardware, allowing for near real-time, in-situ results. In addition, it is highly probable that less costly and more selective IR equipment could be developed for use in monitoring the specific species of interest.

The FT-IR approach, although not fully developed at this time, clearly has some distinct advantages over conventional monitoring methods:

TABLE 7 Comparison of Remote Sensing Extractive Probe Sampling

	Temp. (K)		CO concen. (ppm)		CO_2 concen. (ppm)		THC concen. (ppm)		Soot concen. (ppm)		Efficiency (%)	
Test No.	FT-IR	Probe	FT-IR	Probe	FT-IR	Probe	FT-IR	Probe	FT-IR	Probe	FT-IR	Probe
50	920	416	<20	35	6000	5382	b	17	b	c	>99.67	99.1
53	1400	387	− 27	18	2900	2438	b	7	b	c	99.08	99.0
55	683	344	155	125	1800	1371	b	475	b	c	92.07	69.7
59[a]	2500	421	1350	80	>6000	5965	b	63	b	c	>81.63	97.7

[a]FT-IR in combustion flame, probe above, both above flame in others.
[b]Not measurable by FT-IR.
[c]Not measurable by extractive probe.
Note: Probe data are background-corrected, over the same observation time as the FT-IR values.
Source: McDaniel, 1983.

1. The operation is remote as well as passive. It therefore minimizes potentially objectionable interference with normal plant operation, eliminates the need for personnel in close proximity to a large industrial flare, and makes possible sampling of elevated flares from an off-site location.
2. The approach allows for great flexibility in choosing operational parameters. For example, the spectral resolution, collecting area, spectral range, and measurement time can be easily adjusted in real time to accomodate variations in flare operation, as well as to facilitate both local observations and spatially averaged observations.
3. Monitoring of a number of separated flares can be accomplished in a timely manner from a central measurement facility.
4. FT-IR can accomodate a variety of flare sizes. The threshold sensitivity of FT-IR will in fact improve with size, since for the same spectrometer sensitivity, a longer optical path results in a lower detection threshold.
5. Monitoring equipment can be operated by one person, and there is potential for automated, in-situ data analysis. Therefore, labor costs are minimal, even though initial equipment costs may be relatively high.

Confidence in the results and conclusions derived from FT-IR data is based on the fact that the basic principles and methods are well established, although the specific analysis techniques are not yet perfected for application to flares. The greatest uncertainty in this regard is the lack of an accurate knowledge of the spatial distributions associated with temperature and species concentrations for a "typical" flare. It can be concluded, however, that remote infrared spectroscopy is a viable, and in some ways preferable, means of monitoring flare efficiency.

REFERENCES

Bernstein, L. S. (1980). Band model parameters for the parallel bands of linear triatomic molecules—I. Theory. *J. Quant. Spectrosc. Radiat. Trans.*, *23*: 157.

Clough, S. A., Kneizys, F. X., Rothman, L. S., and Gallery, W. O. (1981). Atmospheric spectral transmittance and radiance: FASCOD1B. *Atmospheric Transmission, 277*, Society of Photographic Instrumentation Engineers, Bellingham, Washington.

Flaud, J. M., and Camy-Peyret, C. (1976). High rovibrational levels of H_2O deduced from high resolution oxygen-hydrogen flame spectra between 2800–6200 cm^{-1}. *Mol. Phys.*, *32*: 499.

Flaud, J. M., Camy-Peyret, C., and Mandin, J. Y. (1977). $H_2{}^{16}O$ hot bands in the 6.3 μm region. *Mol. Phys.*, *34*: 413.

Goody, R. M. (1964). *Atmospheric Radiation, I. Theoretical Basis*, Clarendon Press, Oxford.

Herget, W. F. (1982). Remote and cross-stack measurement of stack gas concentrations using a mobile FT-IR system. *Appl. Opt.*, *21*, No. 4.

Herget, W. F., and Brasher, J. D. (1980). Remote Fourier transform infrared air pollution studies. *Opt. Eng.*, *19*, 508.

Herzberg, Gerhard (1960). *Molecular Spectra and Molecular Structure, II. Infrared and Raman Spectra of Polyatomic Molecules*. D. Van Nostrand Co., Inc., Princeton, New Jersey.

Joseph, D., Lee, J., McKinnon, R., Payne, R., and Pohl, J. (1983). *Evaluation of the Efficiency of Industrial Flares*, U.S. EPA Report No. 600/2-83-070 (NTIS #PB83-263723), Research Triangle Park, North Carolina.

Lindquist, G. H., and Simmons, F. S. (1972). A band model formulation for very nonuniform paths. *J. Quant. Spectrosc. Radiat. Trans.*, *12*: 807.

Ludwig, C. B., Malkmus, W., Reardon, J. E., and Thompson, J. A. L. (1973). *Handbook of Infrared Radiation from Combustion Gases*. NASA Publication SP 3080, Marshall Space Flight Center, Huntsville, Alabama.

Ludwig, C. B., Malkmus, W., and Walker, J. (1981). "The Standard Infrared Radiation Model," AIAA paper 81-1055, Palo Alto, California.

McClatchey, R. A., Benedict, W. S., Clough, S. A., Burch, D. E., Calfee, R. F., Fox, K., Rothman, L. S., and Garing, J. S. (1973). *AFCRL Atmospheric Absorption Line Parameters Compilation*, AFCRL-TR-73-006, Air Force Cambridge Research Laboratories, Bedford, Massachusetts.

McDaniel, M. D. (1983). *Flare Efficiency Study*, US EPA Report No. 600/2-83-052 (NTIS #PB83-261644). Research Triangle Park, North Carolina.

Randall, Charles M. (1974). *Monochromatic Transmittance/Radiance Computations*. TR-0075(5647)-3, The Aerospace Corporation, El Segundo, California.

Young, S. J. (1974). *Band Model Calculations of Atmospheric Transmittance for Hot Gas Line Emission Sources*. TR-0075(5647)-1, The Aerospace Corporation, El Segundo, California.

Young, S. J. (1976). *Band Model Parameters for the 4.3 μm Fundamental Band of CO_2 in the 100–3000 K Temperature Range*. TR-0076(6754-03)-1, The Aerospace Corporation, El Segundo, California.

Young, S. J. (1977). *Description and Use of the Plume Radiation Code ATLES*. SAMSO Report-TR-77-100, Aerospace Corporation, Space and Missile Systems Organization, Air Force Systems Command, Los Angeles, California.

Young, S. J. (1977a). Non-isothermal band model theory. *J. Quant. Spectrosc. Radiant. Trans.*, *18*: 1–28.

5

The Infrared Emission-Absorption Method for Temperature and Species Partial Pressure Determination in Flames and Plumes

CHARLES C. LIMBAUGH *Technology Department, Sverdrup Technology, Inc., AEDC Division, Arnold Air Force Station, Tennessee*

1 INTRODUCTION

The use of the infrared radiative emission of gases to infer physical properties of the gas or fundamental properties of the radiating molecule has long been of utility to the scientific or engineering investigator. In a mixture of gases, for example, the proper selection of frequency bandpass offers the investigator radiation which is a consequence of only one of the species in the mixture, and which, because of the influence of the other species, will reflect the temperature of the mixture as whole. The radiative emission of the gases has long been of vital interest to atmospheric scientists and is of considerable interest for heat transfer applications. One recent dramatic example of the importance of molecular radiation in heat transfer was the identification and subsequent solution of the problem of rocket base heating in the 1960s (Dahm, 1967). This practical engineering problem resulted in investigations into the nature of the radiation from gases typical of combustion products, especially CO_2 and H_2O. Much of the work in this period concerning the description of gaseous radiation is summarized in the comprehensive work by Ludwig et al. (1973).

A consequence of this investigation period was the refinement and development of predictive procedures for the radiant heat transfer from gases that, although based on sound physical principles, were easier and considerably more economical than the more fundamentally correct line-by-line calculations. These predictive techniques, called band models, provided the means to economically calculate radiant emission and transmission of the hot combustion gases using the temperature and pressure of the gas. The inverse problem, in which the radiative properties are used to infer the temperature and pressure of the radiating gas, is a natural extension. Inversion measurements to characterize temperature and species concentration profiles might well be a preliminary to setting up a simpler emission or absorption diagnostic device such as those discussed in previous chapters, since proper interpretation depends on knowledge of those profiles.

This chapter describes an approach to using the band model technology to infer the physical properties of the gas, especially the temperature and species partial pressure. Early work in the highly energetic combustion source of a rocket engine plume showed that the conventional Beer's law approach would not give the correct averaged properties and that a more correct physical model was required. It was apparent that for combustion gases with temperature and pressure gradients, accounting for the inhomogeneities was necessary. This would require a spatial inversion technique such as had been suggested by Pearce (1958) or Freeman and Katz (1960) and was being used with some success to identify the thermal nonequilibrium characteristics of low density plasmas (e.g., McGregor and Brewer, 1966). A description of the state of the art for this inverse problem during that investigative period is provided by Tourin (1966).

Application of these principles for determination of combustion product properties in rockets was first described by Herget (1967). Herget used fixed wavelength measurements of radiance and transmittance with an Abel inversion technique to infer local values of radiance and emissivities. The local temperature was then obtained from Kirchoff's law and pressure from the local emissivity, using published values of the temperature-dependent absorption coefficient. Direct coupling of the band model formulation with a simple radial inversion was subsequently used successfully (Brewer and Limbaugh, 1972).

An alternative approach, in which the spectral signature of the combustion is used to infer the temperature and partial pressure profile, has been described by Krakow (1966), Simmons, Yamada, and Arnold (1969), and more recently by Hommert, Viskanta, and Mellor (1977). This technique, in which spectral data are taken along a single line of sight, has a much simpler data acquisition requirement in that spatial scanning is not required. However, a necessary ingredient for successful execution of the technique is a good a priori knowledge of the shape of the temperature and gas pressure profile along the line of sight. When this is not known, the analysis can become numerically unstable and can either diverge or converge to incorrect values.

The former approach, that of coupling the band model formulation with a geometry model to effect a spatial inversion of the radiative transfer problem for temperature and gas pressure, is the subject of this chapter. For brevity herein, it will be called the infrared emission–absorption (E/A) technique, although in other contexts the term can have a less restrictive meaning. It is hoped this ambiguity will cause no confusion.

It is impossible to provide a description of the E/A technique without considering the band model theory for the prediction of hot gas radiation. There are many excellent works that accomplish this task in depth, and the reader is referred to Goody (1964), Ludwig et al. (1973), and Young (1977) for comprehensive expositions of band model theory.

The ability of the band model formulation to represent the observed spectra from hot H_2O and CO_2 has been examined by a number of investigators, e.g., Lindquist, Arnold, and Spellicy (1975), Young (1977a), Ludwig et al. (1973), and Bernstein, Robertson, and Conant (1980), in a variety of applications. The band models generally provide good modeling of the spectra from these molecules except for a very long air path or in regions of very steep gradient (Lindquist and Simmons, 1972; Young, 1975 and 1975a; Bernstein, Robertson, and Conant, 1980).

Included in the following sections is a brief description of the technical considerations leading to the band model expressions for the optical depth of gases, called the curve of growth. The technique of inversion is then described, followed by discussions of some practical considerations and some recent applications.

2 BAND MODEL THEORY REVIEW

2.1 Radiative Transfer

In a formal model for radiative transfer (Fig. 1), the spectral radiance of a slab of gas of length l with temperature T and pressure p as functions of position may be expressed as

$$N(\bar{\nu},l) = \int_0^l N^0(\bar{\nu},T(s)) \exp\left[-\int_s^l \kappa(\bar{\nu},s')\, ds'\right] \kappa(\bar{\nu},s)\, ds \quad (1)$$

where N^0 is Planck's blackbody function, $\bar{\nu}$ the wavenumber, and κ the pressure- and temperature-dependent absorption coefficient.

Using the Beer–Lambert law and defining the transmittance

$$\tau(s) = \exp\left[-\int_s^l \kappa(\bar{\nu},s')\, ds'\right] \quad (2)$$

as the fractional amount of energy absorbed by the gas along the path, one can write, formally, for the spectral radiance of the gas:

$$N(\bar{\nu},l) = \int_0^l N^0[\bar{\nu},T(s)]\, \frac{\partial \tau}{\partial s}\, ds\,. \quad (3)$$

(This differs from Eq. (14) in Chapter 4 by a minus sign, because of the definition of the transmittance as being from s to l, rather than from 0 to s.) Thus, if one knows the functional form of the absorption coefficient for the gas or the spatial dependence of the transmittance of the gas, the radiative transfer problem for the gas is, in principle, solved.

In practical applications, measurements are made over the finite bandpass of the instrument. The radiance in this case is the result of an average of Eq. (1) or (3) over the instrument

Fig. 1 Homogeneous slab of radiating gas.

bandpass $\Delta\bar{\nu}$, or

$$N_{\Delta\bar{\nu}}(l) = \frac{1}{\Delta\bar{\nu}} \int_{\Delta\bar{\nu}} \int_0^l N^0[\bar{\nu}, T(s)] \frac{\partial\tau}{\partial s}\, ds\, d\bar{\nu} \tag{4}$$

When the $\Delta\bar{\nu}$ is narrow enough that the Planck function is essentially linear over the integral, and expressed at the central value, $\bar{\nu}_0$, the measured radiance can be expressed as

$$N_{\Delta\bar{\nu}}(l) = \int_0^l N^0[\bar{\nu}_0, T(s)] \frac{\partial\bar{\tau}}{\partial s}\, ds \tag{5}$$

which is Eq. (14) in Chapter 4. As before, $\bar{\tau}$ is the average transmittance over the instrument bandpass:

$$\bar{\tau} = \frac{1}{\Delta\bar{\nu}} \int \tau\, d\bar{\nu} \tag{6}$$

Additional formulations for the radiative transfer can be expressed through the use of the absorptance $\alpha = 1 - \tau$ or its integral, the equivalent width W. For a single isolated line, the equivalent width is expressed formally as

$$W = \int_{-\infty}^{\infty} \alpha(\bar{\nu})\, d\bar{\nu} \tag{7}$$

The equivalent width can be interpreted as the spectral width of a line with absorption coefficient of one that gives the same equivalent absorption as the actual spectral line.

With the relation between the transmittance and absorptance, $\alpha = 1 - \tau$, it is seen that

$$\frac{\partial}{\partial s} \int \tau \, d\bar{\nu} = - \frac{\partial W}{\partial s} \tag{8}$$

and Eq. (5) can be expressed in terms of the equivalent width

$$N_{\Delta \bar{\nu}}(l) = - \frac{1}{\Delta \bar{\nu}} \int_0^l N^0[\bar{\nu}_0, T(s)] \frac{\partial W}{\partial s} \, ds \tag{9}$$

When the slab of gas is homogeneous in its physical properties, i.e., there are no gradients of temperature and pressure, the formal expressions, Eqs. (1)–(9), can be further simplified into various forms of the familiar Beer's law and the Kirchoff–Planck law, e.g.,

$$\tau = \exp[-\kappa(\bar{\nu}_0) l] \tag{10}$$

$$N(\bar{\nu}_0, l) = N^0(\bar{\nu}_0, T)(1 - \bar{\tau}) \tag{11}$$

$$N_{\Delta \bar{\nu}}(l) = N^0(\bar{\nu}_0, T)(1 - \bar{\tau}) \tag{12}$$

$$N_{\Delta \bar{\nu}}(l) = N^0(\bar{\nu}_0, T) \frac{W}{\Delta \bar{\nu}} \tag{13}$$

with the instantaneous or average values used depending on the bandpass of the instrument. Thus, in summary, the solution to the radiative transfer problem requires the absorption coefficient [Eq. (1)], transmittance or absorptance [Eq. (3) or (11)], their averaged properties [e.g., Eq. (5) or (12)], or the equivalent width over the bandpass [Eq. (9) or (13)]. The expression to be used depends on the experiment and the spectral bandpass of the instruments.

2.2 Regular Band Models

To illustrate the simplest case of the band model formulation, consider a single, isolated, collision-broadened spectral line (Fig. 2). The absorption coefficient is given by [see Chap. 3, Eq. (36)]

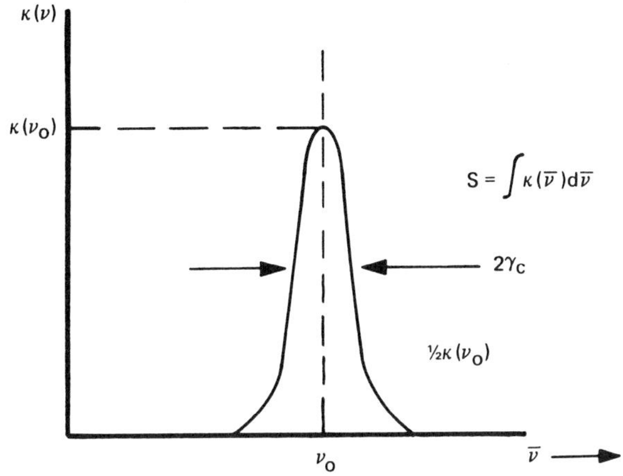

Fig. 2 Absorption coefficient for a single isolated line.

$$\kappa(\bar{\nu}) = \frac{S\gamma_c}{\pi(\bar{\nu}^2 + \gamma_c^2)} \tag{14}$$

in which S is the line strength and γ_c is the collisional half-width at half maximum.

The collision half-width is dependent on collisions of the radiating species with all molecules in the gas as well as the energy states available to the molecule. As discussed in Chap. 3 [Eq. (38)], the form of γ_c for each molecule and process is usually expressed as directly proportional to the pressure and inversely proportional to the temperature of the gas.

$$\gamma_c = \gamma^0 \frac{p}{p_0}\left(\frac{T_0}{T}\right)^N \tag{15}$$

In Eq. (15) p_0, T_0, and γ^0 are values of species pressure, temperature, and collision width at some reference conditions, usually STP.

For a mixture of gases, the contribution from each species in the gas is included in the total linewidth for, say, the jth species, or

$$\gamma_{cj} = \sum_{i=1} \gamma^0_{i,j} \frac{p_i}{p_0} \left(\frac{T_0}{T_i}\right)^{n_{ij}} + \gamma^0_{j,j} \frac{p_j}{p_0} \left(\frac{T_0}{T_j}\right)^{n_{j,j}} \tag{16}$$

where the summation term expresses the contribution to the spectral width by hard sphere type collisions with all molecules while the trailing term expresses the usually more effective resonant broadening from collisions with similar molecules.

Values of $\gamma^0_{i,j}$, $n_{i,j}$, $\gamma^0_{j,j}$, and $n_{j,j}$ may be different for each molecular species in the gas, although $n_{i,j}$ is usually taken to be 0.5 and $n_{j,j}$ is usually taken as 1.0. As described in Chap. 3, experimental values are typically in this range. Ludwig et al. (1973) lists values for several typical combustion products.

The line strength for a gas is similarly expressed as

$$S = S^0 p \frac{T_0}{T} \tag{17}$$

in which the S^0 is the line strength at the reference condition and p is the pressure of the absorbing species. The line strength S^0 has units of frequency, inverse path length and inverse concentration, typically $\mathrm{cm}^{-2}\ \mathrm{atm}^{-1}$. It is a molecular property and is not influenced by the presence of other gases. Through Eq. (17), the absorption coefficient κ includes in its formulation the thermodynamic properties of the gas and the temperature.

Defining the optical path length x,

$$x = \frac{l}{2\pi} \frac{S}{\gamma_c} \tag{18}$$

in which the physical path length is l, the equivalent width can be expressed [Eqs. (2) and (7) and compare Eq. (17) in Chap. 4] as

$$W = \int_{-\infty}^{\infty} \left[1 - \exp\left(\frac{-2x\gamma_c}{\bar{\nu}^2 + \gamma_c^2}\right)\right] d\bar{\nu} \tag{19}$$

or (Ladenburg and Reiche, 1913)

$$W = 2\pi\gamma_c x \exp(-x)[I_0(x) + I_1(x)] \tag{20}$$

where the $I_0(x)$ and $I_1(x)$ are Bessel functions. Defining

$$f(x) = x \exp(-x)[I_0(x) + I_1(x)] \tag{21}$$

then simply,

$$W = 2\pi\gamma_c f(x). \tag{22}$$

This $f(x)$ is called the Ladenburg–Reiche function and is tabulated (Kaplan and Eggers, 1956; or Ludwig et al., 1973). Equation (22), expressing the relationship between the equivalent width of the spectral line and the gas parameters (x) is called the curve of growth and completes the band model formulation for the isolated spectral line.

Measurements of transmittance or absorptance (equivalent width) and radiance of the isolated spectral line from such a homogeneous slab of gas are related functionally to the temperature and pressure by the x parameter in the curve of growth and the Kirchoff–Planck law, or Eq. (12), and

$$W = 2\pi\gamma_c f(x) = (1 - \bar{\tau})\, \Delta\bar{\nu} \tag{23}$$

which can be solved for the temperature and pressure.

The absorption coefficient for single isolated spectral lines is known and can be described by either the Lorentz, Doppler, or Voigt profiles (e.g., Penner, 1959). Other line broadening mechanisms, e.g., natural, are usually insignificant at the temperatures and pressures typical of combustion gases.

Molecular spectra can be very complex, however, and for many molecules, isolating a single line for subsequent measurement and analysis can be difficult even at high resolution. Additionally, many spectral measurements are made at low or moderate resolution, and the instrument bandpass can encompass a very large number of overlapping spectral lines, which similarly complicates the evaluation. In this case, a band model based on a single, isolated spectral line is inappropriate, and more complete considerations must be made.

As a first approximation to this more complex case, consider a model proposed by Elsasser (1942) in which the spectrum is composed of a series of regularly occurring, equal intensity lines (Fig. 3). In this case $\kappa(\bar{\nu})$ is given by the contribution from each line

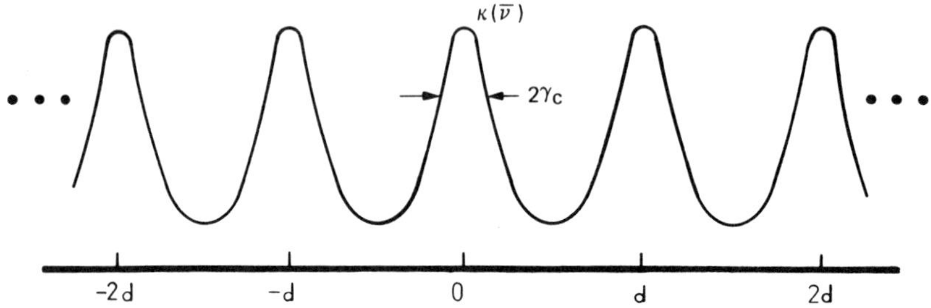

Fig. 3 Series of regularly spaced equal intensity lines for Elsasser model.

$$\kappa(\bar{\nu}) = \sum_{n=-\infty}^{\infty} \frac{S\gamma_c}{\pi[(\nu - nd)^2 + \gamma_c^2]} \tag{24}$$

with the resultant (see for example Goody, 1964) average equivalent width

$$\frac{\bar{W}}{d} = 1 - \frac{1}{2\pi} \int_{-\pi}^{\pi} \exp\left\{- \frac{S}{d}\, pl\, \frac{\sinh[2\pi(\gamma_c/d)]}{\cosh[2\pi(\gamma_c/d)] - \cos z}\right\} dz \tag{25}$$

Although the integral does not have a closed form solution, it can be evaluated numerically or from published tables (Goody, 1964; or Ludwig et al., 1973).

The model used for the development of this Elsasser model, equally spaced lines of equal intensity, is in fact often approximated in nature by linear and diatomic molecules, and for such molecules can provide an adequate model for the radiative transfer.

2.3 Random Band Models

The more complex polyatomic molecules, e.g., H_2O and CO_2, demonstrate a considerably more complex spectrum, and the physical model imposed by using the Elsasser band model is not approximated. For example, Fig. 4 (Krakow, Kiech, and McAdoo, 1977) is the high resolution ($0.2\ cm^{-1}$) spectrum of a propane torch which shows the very complex spectra that can arise in the H_2O and CO_2 bands. Indeed, in such complicated spectra, the

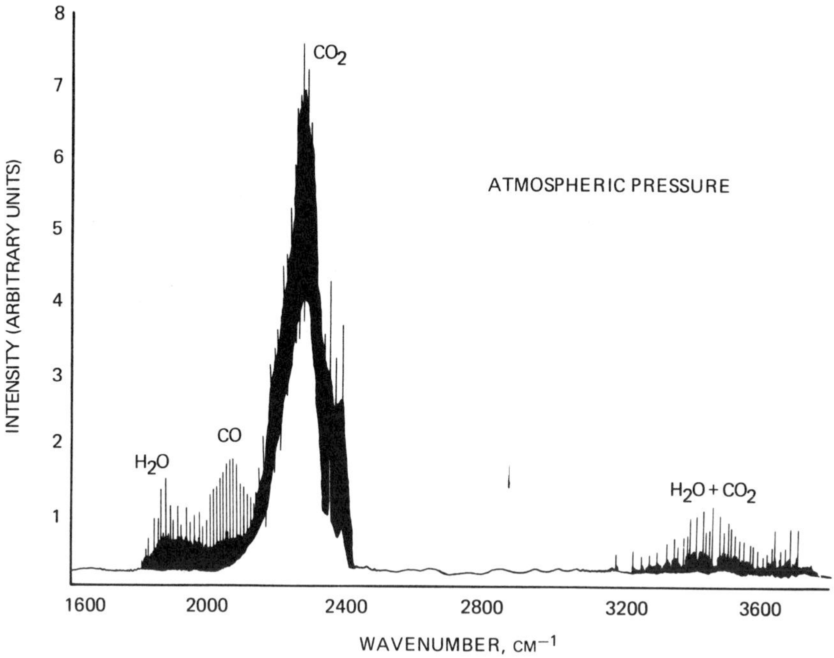

Fig. 4 Emission spectrum of propane torch, 1800 to 3700 cm^{-1}, 0.2-cm^{-1} resolution (from Krakow, Kiech, and McAdoo, 1977).

irregularly spaced line positions and intensities appear random, and a statistical model for the spacings and strengths is suggested. There are several of these random band models in which the line strength is described analytically by a probability distribution function and the line position is handled by using an average line spacing in the interval d. The shape of each spectral line is assumed to be the same, only the strength is varied according to the probability distribution function. Goody (1964), Ludwig et al. (1973), and Young (1975, 1975a, 1977, 1977a) provide the reader an excellent background on the development of the random band model theory.

In the case in which each line has the same shape but arbitrary strength and in which there are many such randomly distributed lines occurring within the bandpass $\Delta\bar{\nu}$, it has been shown (Goody, 1964, 1964a; or more recently Koo et al., 1979) that the transmittance over an interval can be expressed in terms of the contribution of each spectral line to the resultant average transmittance

$$\bar{\tau} = \prod_{i=1}^{n}\left(1 - \frac{W_i}{\Delta \bar{\nu}}\right) \tag{26}$$

in which W_i is the equivalent width of each individual spectral line.

In the limit of large n, we obtain Eq. (14) in Chap. 4,

$$\bar{\tau} = \exp(-\bar{W}/d) \tag{27}$$

where $\bar{W}$ is the average equivalent width over the interval and $d = \Delta\bar{\nu}/n$ is the average spacing of the lines in the interval. The average $\bar{W}$ can be expressed in terms of the distribution function [as in Eq. (18) in Chap. 4]:

$$\bar{W} = \int_0^\infty \mathscr{P}(S)W(S)\,dS \tag{28}$$

where $\mathscr{P}(S)$ is the strength probability distribution function and is normalized:

$$\int_0^\infty \mathscr{P}(S)\,dS = 1 \tag{29}$$

The average equivalent width expression can be put into the same functional form as that for the single isolated spectral line,

$$\ln \bar{\tau} = -\frac{\bar{W}}{d} = -2\pi\frac{\gamma}{d}f(x) \tag{30}$$

where γ is the half-linewidth at half maximum and d is the average distance between spectral lines. The x parameter in Eq. (30) is defined similarly to the x of Eq. (18) for the band model for a single isolated line, using averaged parameters

$$x = \frac{l}{2\pi}\frac{S/d}{\gamma/d} \tag{31}$$

in which l is the path length through the absorbing medium. The S/d is obtained from

$$\frac{S}{d} = \bar{K}_0 p \tag{32}$$

in which p is the species pressure and $\bar{K}_0$ is the temperature- and wavenumber-dependent average absorption coefficient band model parameter over the interval $\Delta \bar{\nu}$. The γ/d is the product of the half-width at half maximum γ and average reciprocal line spacing band model parameter $1/d$.

The band model parameters $\bar{K}_0$ and $1/d$ are obtained from the literature (Ferriso, Ludwig, and Thomson, 1966; Ludwig et al., 1973; Young, 1976, 1976a). Care in using them must be exercised, however, since some sources (e.g., Ludwig et al., 1973) list an absorption coefficient $\bar{K}^0$ which is referenced to STP, and in this case

$$\frac{S}{d} = \bar{K}^0 p \frac{T_0}{T} \tag{33}$$

whereas Young (1976, 1976a, 1977a) incorporates the temperature dependence into the listing, and Eq. (32) can be used directly. The available band model coefficients for H_2O vapor are derived from a large number of measurements reported by Ferriso, Ludwig and Thomson (1966), while the CO_2 band model parameters are obtained from line-by-line calculations (most recently Bernstein, 1980), although experimental determinations have been similarly reported (Kunitomo and Osumi, 1975). Band model parameters and their sources are also available in the literature for other molecules (e.g., Ludwig et al., 1973).

The linewidth γ depends on the physical line-broadening mechanism: collisional or Doppler; the functional form of $f(x)$ depends on both the probability distribution function $\mathscr{P}(S)$ and the broadening mechanism. For example, the absorption coefficient for a Doppler-broadened line is given by

$$\kappa_D = \frac{S}{\gamma_D} \sqrt{\frac{\ln 2}{\pi}} \exp\left(\frac{-\ln 2(\bar{\nu} - \bar{\nu}_0)^2}{\gamma_D^2}\right) \tag{34}$$

and the Voigt profile for combined collisional and Doppler broadening

$$\kappa_{\bar{\nu}} = \frac{S\sqrt{\ln 2}}{\pi\gamma_D} \frac{\gamma_c}{\gamma_D} \sqrt{\frac{\ln 2}{\pi}} \int_{-\infty}^{\infty} \frac{\exp(-y^2)}{a^2 + (\xi - y)^2}\, dy \qquad (35)$$

where m is the molecular mass, c the speed of light, $\bar{\nu}_0$ the wavenumber at line center, and

$$\gamma_D = \sqrt{\frac{2 \ln 2\, kT}{mc^2}}\, \bar{\nu}_0 \qquad (36)$$

$$a = \sqrt{\ln 2}\, \frac{\gamma_c}{\gamma_D} \qquad (37)$$

$$\xi = \sqrt{\ln 2}\, \frac{\bar{\nu} - \bar{\nu}_0}{\gamma_D} \qquad (38)$$

for comparison to the purely collision-broadened absorption coefficient, Eq. (14).

Table 1 lists the curves of growth associated with the most commonly used probability distribution functions for the line strengths for both collision- and Doppler-dominated broadening. The reader is referred to Ludwig et al. (1973) or Young (1975, 1975a, 1977) and their references for in-depth discussion of the considerations leading to each of the expressions. Asymptotic forms and simplified expressions for evaluating the integrals are given. In the situation in which neither collisional nor Doppler broadening is the dominant physical mechanism, Rodgers and Williams (1974) and Young (1978) show that a proper method of combining the Lorentzian and Doppler curves of growth is given by

$$\left(\frac{W}{d}\right)_V = \left\{\left(\frac{W}{d}\right)_c^2 + \left(\frac{W}{d}\right)_D^2 - \left[\frac{(W/d)_c (W/d)_D}{2\pi(\gamma_c/d)x_c}\right]^2\right\}^{1/2} \qquad (39)$$

Here x_c is that x defined in Eq. (31) with γ being γ_c. The entire denominator equals (Sl/d), the weak line limit for (W/d) for a line of any shape. Weak and strong line limits will be discussed further in the following section. Ludwig et al. (1973) give alternate approximating expressions.

TABLE 1 Probability Distribution Functions and Corresponding Curves of Growth for Random Band Model

	$\ln \tau = -\overline{W}/d$	
P(S)	Collision-broadened	Doppler-broadened
$\delta(S - \bar{S})$	$2\pi \frac{\gamma_c}{d} f(x_c)$	$\sqrt{\frac{\pi}{\ln 2}} \frac{\gamma_D}{d} g(x_D)$
$\bar{S}^{-1} e^{-S/\bar{S}}$	$2\pi \frac{\gamma_c}{d} x_c \left[1 + \frac{\pi}{2} x_c\right]^{-1/2}$	$\sqrt{\frac{\pi}{\ln 2}} \frac{\gamma_D}{d} \int_0^\infty e^{-z} g(x)_D z)\ dz$
$S^{-1} e^{-S/\bar{S}}$	$2 \frac{\gamma_c}{d} [(1 + 2\pi x_c)^{1/2} - 1]$	$\sqrt{\frac{\pi}{\ln 2}} \frac{\gamma_D}{d} \frac{2}{\sqrt{\pi}} \int_0^\infty \ln[1 + x_D e^{-z^2}]\ dz$

where

$f(x) = xe^{-x}[I_0(x) + I_1(x)]$ $\quad$ $g(x_D) = \frac{2}{\sqrt{\pi}} \int_0^\infty (1 - e^{-x_D e^{-z^2}})\ dz$

$\bar{S}$ = Mean line strength

$x_c = \frac{l}{2\pi} \frac{S/d}{\gamma_c/d}$ $\quad$ $x_D = \frac{l}{2\pi} \frac{S/d}{\gamma_D/d}$

Source: Ludwig et al. (1973) and Young (1975).

2.4 Inhomogeneous Paths

When the gas in the path is inhomogeneous in temperature and/or pressure of the absorbing species, the integrals of Sec. 2.1 must be accomplished to evaluate the radiative transfer. Thus the simplification leading to Eq. (23) is not valid, and a more complex formulation must be pursued. With instrumentation of finite bandpass, the integration is further complicated by the temperature variation of the absorption coefficient at the different frequencies of the bandpass.

Direct evaluation of the space-dependent integral [e.g., Eq. (4)] at each frequency in the bandpass of interest can lead to extensive and expensive computational effort. Separation of the frequency and spatial integrations so that the transmittance over a wider frequency interval is represented by the band model theory is an approximation that is often used to reduce this effort. A generally useful model to further reduce the computation effort for the inhomogeneous path is to replace the inhomogeneous path by an equivalent path with the same transmittance over the bandpass of interest.

This approach, called the Curtis–Godson approximation, is not, however, valid in all applications. Young (1975, 1975a, 1979) examines the Lindquist–Simmons approximation (Lindquist and Simmons, 1972) and determines the appropriate conditions and resulting formulation for those special cases for which the Curtis–Godson approximation fails. Each approach is discussed briefly below.

Curtis–Godson Approximation

The Curtis–Godson approximation gives a simplification arising from the requirement that each of the curves of growth be asymptotic to a physical condition at the limits of small and large x. At small x, the so-called weak line limit holds, $f(x) \approx x$, and the average equivalent width grows linearly with path length. This is usually taken to occur for $x \leq 0.1$. At large x, the so-called strong line limit or square root region, $f(x) \approx \sqrt{2x/\pi}$. This is usually taken for $x \geq 10$.

Thus, for any gas path, including inhomogeneous paths, at the weak line limit:

$$\ln \bar{\tau} = -\frac{Sl}{d} \tag{40}$$

and, at the strong line limit:

$$\ln \bar{\tau} = -2 \sqrt{l \frac{S}{d} \frac{\gamma}{d}} \tag{41}$$

or, defining

$$L \frac{\mathsf{S}}{d} = \int_0^L \frac{S}{d} \, ds \tag{42}$$

and

$$\frac{\Gamma}{d} L \frac{\mathsf{S}}{d} = \int_0^L \frac{\gamma}{d} \frac{S}{d} \, ds \tag{43}$$

where L is the total length of the inhomogeneous path, one has the necessary asymptotic limits for any curve of growth defined in terms of the path-averaged band model coefficients, S/d and Γ/d. Using these averaged parameters, the transmittance for an inhomogeneous path can be expressed as

$$\ln \bar{\tau} = -2\pi \frac{\Gamma}{d} f(\mathsf{X}) \tag{44}$$

where

$$\mathsf{X} = \frac{L}{2\pi} \frac{\mathsf{S}/d}{\Gamma/d} \tag{45}$$

and the path-averaged band model parameters were defined in Eqs. (42) and (43). Equations (42) and (43) comprise the Curtis–Godson approximation. The principal advantage to the Curtis–Godson approximation has been to provide an easy means for determining the space-averaged band model parameters. For, if one models the continuously inhomogeneous path by a series of small homogeneous regions (Fig. 5), each having path length l_i, the space-dependent integrals of the radiative transfer problem and the Curtis–Godson approximation can be expressed as

$$N_{\Delta\bar{\nu}}(L) = \sum_i N^0(\bar{\nu}_0, T_i)(\bar{\tau}_{i-1} - \bar{\tau}_i) \tag{46}$$

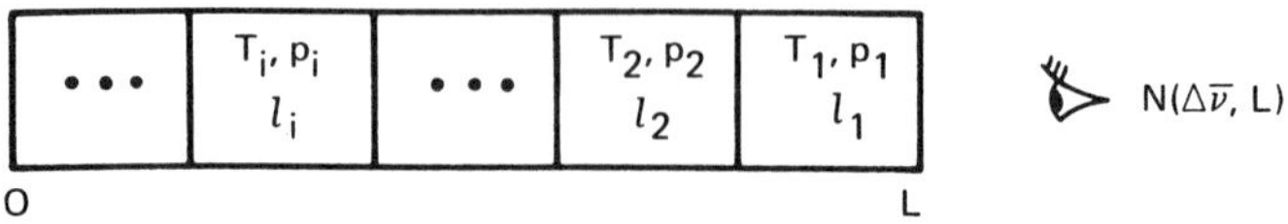

Fig. 5 Model for inhomogeneous slab of radiating gas.

$$\ln \bar{\tau} = -2\pi \frac{\Gamma}{d} f(X) \tag{47}$$

$$L \frac{S}{d} = \sum_i (S/d)_i l_i \tag{48}$$

$$\frac{\Gamma}{d} = \frac{\sum_i (S/d)_i l_i (\gamma/d)_i}{\sum_i (S/d)_i l_i} \tag{49}$$

The above intuitive development of the Curtis–Godson approximation is essentially that suggested by Curtis (1952) and Godson (1953) and followed by most investigators since (e.g., Kaplan, 1959). Armstrong (1968) examines the approximation from a more formal approach.

The Curtis–Godson approximation has been scrutinized over the years, and comparisons between "exact" and approximate computations have been made (Walshaw and Rogers, 1963; or Goody, 1964a). Krakow et al. (1966) report measurements made at several laboratories which provide a means to directly examine the validity of the Curtis–Godson approximation. Their conclusions were that for the conditions of those experiments, the Curtis–Godson approximation gives the correct results to an accuracy better than that of the measurements.

Lindquist–Simmons Approximation

In other considerations, Lindquist and Simmons (1972) and Young (1975, 1975a, 1977, 1977a) have shown the Curtis–Godson approximation can give erroneous results in special cases (regions of very steep spatial gradients or a hot source viewed through a long path length intervening atmosphere). They appropriately argue that, since the derivative of the transmittance occurs in the radiance equation, better accuracy for the radiance is obtained by approximating the derivative functions. These considerations lead to the expression

$$\frac{1}{d}\frac{d\overline{W}}{ds} = \frac{S}{d}\, y(x_e, \gamma/\gamma_e) \,. \tag{50}$$

Here γ/γ_e is an inhomogeneity factor expressing the ratio of the local value of the linewidth to the path-averaged linewidth to that point

$$\gamma/\gamma_e = \frac{(\gamma/d)\int \overline{K}_0 p \; ds}{\int \overline{K}_0 p(\gamma/d) \; ds} \tag{51}$$

The absorption coefficient and linewidth are evaluated using formulas appropriate to the physical broadening coefficient present. Then, x_e is a measure of the optical path length given by

$$x_e = \frac{\int S \; ds}{2\pi\gamma_e} = \frac{\left(\int \overline{K}_0 p \; ds\right)^2}{2\pi\int \overline{K}_0 p(\gamma/d) \; ds} \tag{52}$$

We have added the subscript e to denote the path averaging which gives rise to "equivalent" quantities and to distinguish against the x defined in Eqs. (31); Young (1975) and Lindquist and Simmons (1972) simply use x. The $y(x_e, \gamma/\gamma_e)$ function is shown in their work to be physically constrained such that

$$0 \leqslant y(x_e, \gamma/\gamma_e) \leqslant 1 \tag{53}$$

for all x_e and γ/γ_e. It assumes varied functional forms depending on the broadening mechanisms and the probability distribution function. In the following discussion, we denote γ/γ_e by ζ, where Young used ρ.

Young (1975) and Lindquist and Simmons (1972) find that the condition imposed in Eq. (53) is violated by the Curtis–Godson approximation when the inhomogeneity factor ζ is very large ($\zeta \gg 2$) and x_e is small or medium ($x_e < 10$) near the beginning of the optical path. As path length increases, x_e becomes larger and the combined requirement of $\zeta \gg 2$ and $x_e < 10$ is less likely to be met. This situation is most likely to occur in calculations of radiative transfer from a hot flame through a cool atmosphere (large inhomogeneity) and in the band wings when the absorption coefficient $\overline{K}_0$, and consequently x_e, is small. Other environments which have very high temperature or pressure gradients can similarly give large values of this inhomogeneity factor.

With Eq. (50), the solution to the radiative transfer problem becomes

$$N_{\Delta\bar{\nu}}(L) = \int_0^L N^0(\bar{\nu}_0, T)\, \frac{S}{d}\, y(x_e, \zeta)\, \tau\, ds \tag{54}$$

$$\ln \tau = -\int_0^L \frac{S}{d}\, y(x_e, \zeta)\, ds \tag{55}$$

Here S, x_e, and ζ are functions of the path coordinate s, the latter through their integration limits. The integrals in Eqs. (54) and (55) are evaluated numerically.

Table 2 lists the functional forms for y for the various probability distribution functions listed in Table 1.

3 RADIAL INVERSION

The band model formulation of the radiative transfer problem provides a pair of coupled nonlinear equations relating the temperature and partial pressure of a molecular species to the radiance and transmittance of a slab of hot gas. When the gas contains spatial gradients of temperature and pressure, as is often the case for large energetic combustion sources such as the exhaust of a jet engine or rocket, the Curtis–Godson approximation or the Lindquist–Simmons approximation of the previous section is used to describe the functional relationship between the radiance and transmittance and the gas profiles of temperature and pressure. Thus, if one knows the radiance and transmittance of the slab of hot gas, in principle one can determine the pressure and temperature of the gas.

In this more complex case it is necessary to assume or specify a geometry model and acquire data adequate for specifying all the parameters in the geometry model. The specific geometry model used is dictated by the geometrical characteristics of the hot gases for which the temperature and species partial pressures are required. When the geometry is cylindrical, radial inversion techniques are appropriate. Their use to determine the spatial profile of physical parameters was well established in the early 1960s with plasma and arc jet research. That work was usually concerned with evaluation of the inversion of Abel's equations and was attacked by a variety of methods (e.g., Pearce, 1958; Freeman and Katz, 1960; Barr, 1962; Cremers and

TABLE 2 Derivative Functions for Random Band Model Probability Distributions, Lindquist–Simmons Approximation

$\mathscr{P}(s)$	$\frac{d}{ds}\frac{\bar{w}}{d} = \frac{S}{d}\, y(x_e, \zeta)$ Collision-broadened	Doppler-broadened
$\delta(S - \bar{S})$	$\frac{2\zeta}{\pi}\int_0^{\pi} H(\zeta,\theta)e^{-x_e(1+\cos d\theta)}$	$\frac{2}{\sqrt{\pi}}\int_0^{\infty} e^{-z^2 - x_e e^{-\zeta^2 z^2}}\, dz$
$S^{-1}e^{-S/\bar{S}}$	$\frac{2\zeta}{\pi}\int_0^{\pi} H(\zeta,\theta)G(x_e\theta)\; d\theta$	$\frac{2}{\sqrt{\pi}}\int_0^{\infty} \frac{e^{-z^2}}{(1 + x_e e^{-\zeta^2 z^2})^2}\, dz$
$S^{-1}e^{-S/\bar{S}}$	$\frac{2\zeta}{\pi}\int_0^{\pi} H(\zeta,\theta)\sqrt{G(x_e,\theta)}\; d\theta$	$\frac{2}{\sqrt{\pi}}\int_0^{\infty} \frac{e^{-z^2}}{1 + x_e e^{-\zeta^2 z^2}}\, dz$

where

$$H(\zeta,\theta) = \frac{1}{(\zeta^2 + 1) + (\zeta^2 - 1)\cos\theta} \qquad x_e = \frac{[\int \bar{K}_0 p\; ds]^2}{2\pi \int \bar{K}_0 p(\gamma/d)\; ds}$$

$$G(x_e,\theta) = \frac{1}{[1 + x_e(1 + \cos\theta)]^2} \qquad \zeta = \frac{(\gamma/d)\int \bar{K}_0 p\; ds}{\int \bar{K}_0 p(\gamma/d)\; ds}$$

Source: Lindquist and Simmons (1972) and Young (1975, 1977).

Birkebak, 1966; Shelby, 1976). These techniques have generally become called onion peeling (Ludwig et al., 1973) because the determination of a profile commences at the outer edge and proceeds to the center, as in stripping off the shells of an onion, consecutively revealing the next inner shell.

A general feature of inversion problems, as discovered by the early investigators, was the sensitivity of results to the uncertainties in the data, and since physical data almost invariably have uncertainties or noise associated with them, data smoothing is tantamount to successful inversion, either by Abel inversion or other techniques. Barr (1962) and Cremers and Birkebak (1966) examined and subsequently recommended least squares smoothing of the data to a polynomial as did Blair (1974). Blair used a single polynomial of a low degree while Cremers and Birkebak used different polynomials over different segments of data. Barr used quadratic functions for least squares fitting over a limited range of data, but did not require smoothness at each interval boundary. More modern investigators are incorporating spline functions with least squares techniques (e.g., Glasser, Chapelle, and Boettner, 1978; Balasubramanian, Norrie, and DeVries, 1979; Deutsch and Beniaminy, 1983), although Shelby (1976) or equivalently Shelby and Limbaugh (1977), and Limbaugh (1977) appear to be the earliest application to actual uncertain data. Shelby (1976) presents the matrix equations to accomplish the least squares fitting to cubic splines as well as the propagation of data uncertainties through the curve fit and subsequent Abel inversion.

Hughey and Santavicca (1982) recently compared the results of three different inversion techniques: finite width onion peel, Abel transformation (infinitesimal width onion peel), and Fourier transformation. They were applied to computer-simulated absorption measurements for which the property profile functional form was known. The results showed that the presence of measurement error coupled with a small number of points used in the inversion can lead to significant errors in the results. Indeed, in effecting radial inversion, one should use a data spatial resolution consistent with that of the measurement instrument. With physical data, which will be inherently uncertain, a large number of measurements will serve to provide good estimates of average values, and provide better estimates of least squares derived functions.

Determination of profiles in an asymmetric medium has long been of interest because such a capability will remove the geometry assumption requirement and generalize the techniques and applications. Wang (1976, 1977) proposed one approach for accomplishing this, requiring measurements at several angles through the medium. Ray and Semerjian (1983) recently reported the application of techniques similar to those used with CAT scanners in medical technology to measurements in an artificially seeded flat flame burner with good results. Although the laser tomography has apparently not as yet been applied outside the laboratory, it is an extremely promising approach and should be watched carefully for further developments.

Past applications of the E/A techniques have been to jet engine or rocket exhausts which are generally adequately modeled by a symmetric medium such as a series of concentric cylinders, each of which is homogeneous in temperature and pressure.

Figure 6 shows one way in which an E/A experiment for such a combustion souce can be accomplished. The graybody source on one side of the plume provides a hot continuum source for transmission measurements, and a detector optically aligned with the graybody provides for the measurements of the radiation. Figure 7 shows a typical instrumentation block diagram. Filters at the detector provide a choice of radiation bandpass. By using mechanical choppers at significantly different frequencies at the

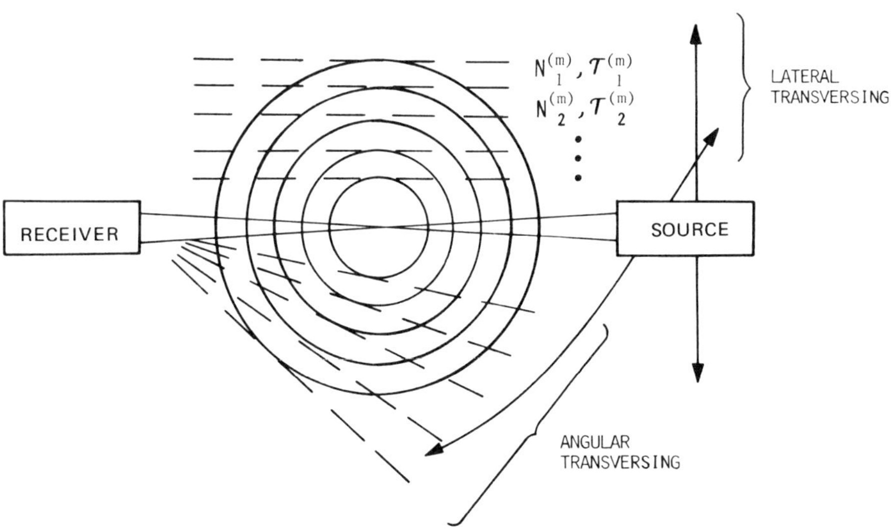

Fig. 6 Typical E/A plume traversing techniques.

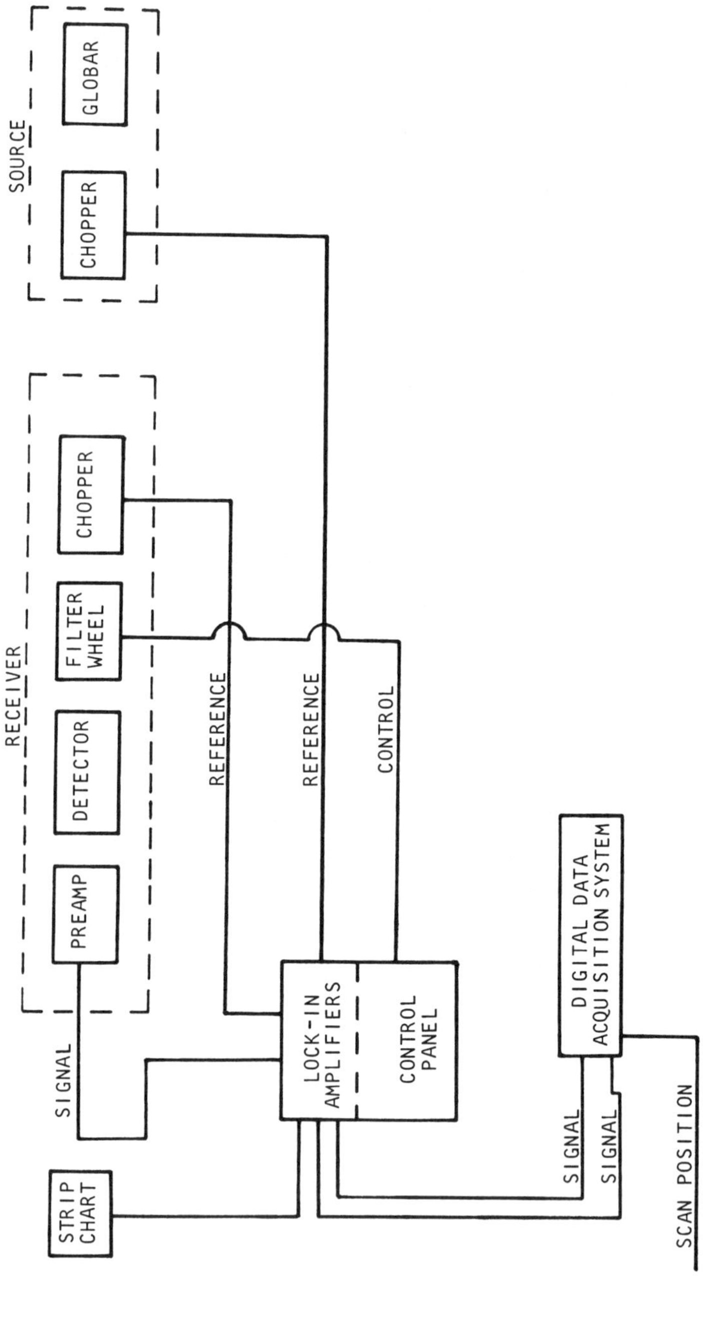

Fig. 7 Block diagram of typical E/A instrumentation.

detector and graybody source, phase-locked amplifiers allow one to record either plume radiation alone or plume and graybody radiation. Determination of the graybody signal with no emitting–absorbing gas in the optical path provides the 100% transmission signal. Consecutive measurements of radiance and transmittance across the plume from the outer edge through the center provide lateral profiles of radiance and transmittance. The traversing can be accomplished either by lateral or angular traversing, as shown in Fig. 6.

By way of illustration, Fig. 8 (Limbaugh and Kneile, 1984) shows raw data smoothed profiles or radiance and transmittance in the 4.3-μm band of CO_2 which were obtained from a full-size rocket in a simulated altitude test. The example profiles shown in Fig. 8 show the normal experimental noise characteristic of data and include more than 3000 independent measurements of radiance and transmittance. The E/A technique can be sensitive to such noise, since it is possible for the normal random experimental uncertainties to produce a radiance and transmittance pair that is physically nonsensical; consequently, the data can be smoothed by any of a number of least squares techniques.

There have been two approaches (Brewer and Limbaugh, 1972, and Young, 1981) used to effect inversion of the spatial integrals to determine the radial profiles for temperature and species partial pressure. Each are discussed in turn below. As developed and applied, both depend on a cylindrical geometry for the plume. The method of Brewer and Limbaugh uses the geometry only for the calculation of the optical path along the line of sight and, in principle, is immediately adaptable to other geometries.

3.1 Finite Width Onion Peel Inversion

In the inversion procedure used by Brewer and Limbaugh (1972) and Brewer (1975), the smoothed data are used to determine data at each of several positions across the plume. Each consequent data point defines the radiance and transmittance of a zonal slice across the plume. The number of zones and their radial widths are arbitrary but specifically defined for a given inversion. The zone widths do not have to be uniform in size but are generally taken so. As shown in Fig. 6, there is an optical path associated with each homogeneous zone k, and a measured radiance $N_k^{(m)}$ and transmittance $\tau_k^{(m)}$ for each optical path. The portion of the optical path between the plume boundary and the instrument is assumed to have 100 percent transmittance and zero radiance. When there is measurable absorption or radiance by the environ-

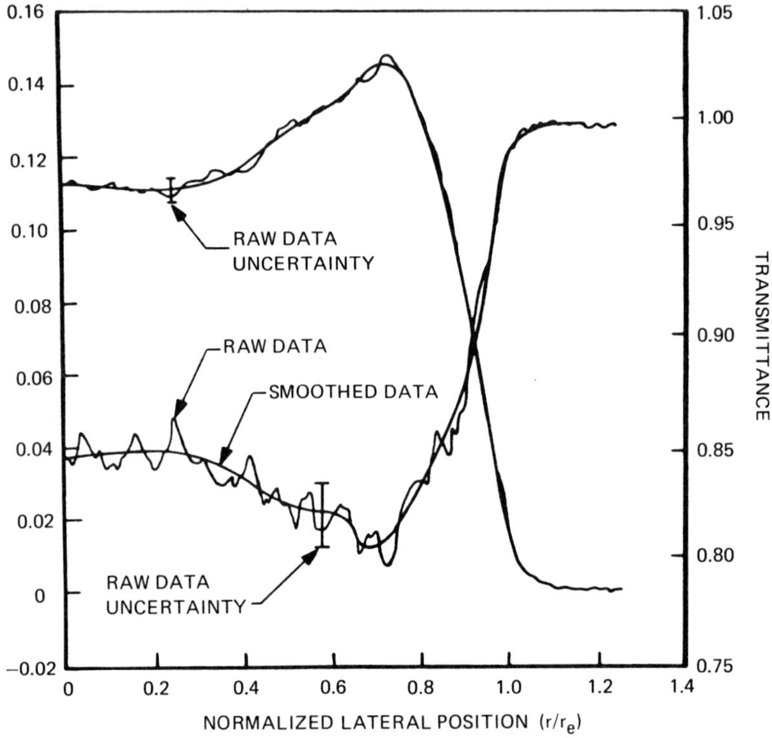

Fig. 8 Example of CO_2 radiance and transmittance profiles from rocket exhaust. Reprinted with permission from the *Journal of Quantitative Spectroscopy and Radiative Transfer 31*: 161–171, Limbaugh and Kneile, "Uncertainties Propagation for Combustion Diagnostics Using Infrared Band Models," Copyright 1984, Pergamon Press, Ltd.

ment outside the actual plume, the geometrical model can expand to the instrument face and thus include this region. Consequently, the effect of any recirculating plume or atmospheric gases can be included in the analysis.

As described by Brewer and Limbaugh (1972) and elsewhere, application of the technique is accomplished in the manner typical of a multilayer inversion problem. A measurement pair $(N_1^{(m)}, \tau_1^{(m)})$ at the outer edge of the plume, used with Eqs. (45)–(49), gives

$$N_1^{(m)} = N^0(T_1)(1 - \tau_1^{(m)}) \tag{56}$$

$$\ln \tau_1^{(m)} = -2\pi\left(\frac{\Gamma}{d}\right)_1 f(X_1) \tag{57}$$

where

$$X_1 = \frac{l_1}{2\pi}\frac{(S/d)_1}{(\Gamma/d)_1} \tag{58}$$

the subscript 1 denotes the first measurement at the outer edge of the plume, and the superscript "m" denotes that the quantity results from the measurement process. The gas properties, temperature and partial pressure, appear in the blackbody function N^0 and in the S/d and Γ/d parameters. The l_1 is the path length through the absorbing material. The $f(x)$ is chosen appropriately from Table 1 or other sources. The temperature T_1 is the only unknown in Eq. (56) and can be solved directly from blackbody tables with

$$N^0(T_1) = \frac{N_1^{(m)}}{1 - \tau_1^{(m)}} \tag{59}$$

or analytically using the Planck blackbody function.

The absorbing species partial pressure is the only remaining unknown and can be found numerically by determining the root to

$$\ln \tau_1^{(m)} + 2\pi\left(\frac{\Gamma}{d}\right)_1 f(X_1) = 0 \tag{60}$$

The iteration of Eq. (60) generally converges rapidly, and the bisection method is appropriate.

The measurement pair $(N_2^{(m)}, \tau_2^{(m)})$ is then used with application of Eqs. (46) and (47) to determine (T_2, p_2), the unknown properties of the central zone of the second slice across the plume. The (T_1, p_1) and (T_2, p_2) are then used in conjunction with the measurement $(N_3^{(m)}, \tau_3^{(m)})$ to determine (T_3, p_3), and the process is continued to the center of the plume in the fashion typical of an "onion peel" inversion.

To illustrate this application to the kth slice, consider Fig. 9, for which $(N_k^{(m)}, \tau_k^{(m)})$ are the measured radiance and transmittance, the $(T_{k,i}, p_{k,i})$ for all $i < k$ being known. A double subscript notation is used to denote the properties of a homogeneous zone across the plume. The leading subscript k identifies the slice or optical path across the plume, while the trailing subscript identifies the homogeneous zone along the line of sight. The trailing subscript starts with one on the observer side of the plume and goes to $2k-1$ on the source side of the plume, for cylindrical geometry. With this double subscript notation, the measured transmittance $\tau_k^{(m)}$ and the calculated $\tau_{k,2k-1}$ are the same physical quantity. The band model formulation in this notation for this measurement yields

$$N_k^{(m)} = \sum_{i=1}^{2k-1} N^0(T_{k,i})(\bar{\tau}_{k,i-1} - \bar{\tau}_{k,i}) \tag{61}$$

$$\ln \bar{\tau}_{k,i} = 2\pi\left(\frac{\Gamma}{d}\right)_{k,i} f(X_{k,i}) \tag{62}$$

$$X_{k,i} = \frac{L(S/d)_{k,i}}{2\pi(\Gamma/d)_{k,i}} \tag{63}$$

$$L\left(\frac{S}{d}\right)_{k,i} = \sum_{j=1}^{i} l_{k,j}\left(\frac{S}{d}\right)_{k,j} \tag{64}$$

$$\left(\frac{\Gamma}{d}\right)_{k,i} = \frac{\sum_{j=1}^{i} l_{k,j}(S/d)_{k,j}(\gamma/d)_{k,j}}{L(S/d)_{k,i}} \tag{65}$$

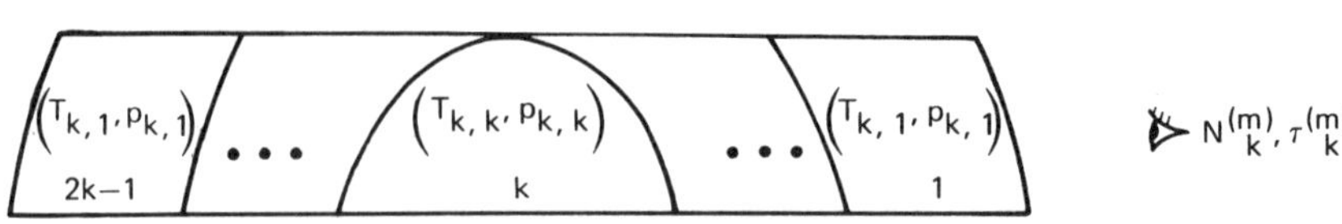

Fig. 9 Diagram of cylindrically symmetric zones contributing to the kth measurement of radiance and transmittance.

Equations (61) and (62) are nonlinear and coupled so that effective solution is best carried out numerically.

To illustrate a formulation for solution of the unknown $(T_{k,k}, p_{k,k})$, note that the right side of Eqs. (61) and (62) can be expressed functionally:

$$N_k = N_k(T_{k,k}, p_{k,k}) \tag{66}$$

$$\tau_k = \tau_k(T_{k,k}, p_{k,k}) \tag{67}$$

in which all other temperatures and pressures can be considered known and constant.

The objective is to determine $(T_{k,k}, p_{k,k})$ such that the analytical evaluation of Eqs. (66) and (67) gives the same value as the measurement, which can be written formally as

$$N_k^{(m)} - N_k(T_{k,k}, p_{k,k}) = 0 \tag{68}$$

$$\tau_k^{(m)} - \tau_k(T_{k,k}, p_{k,k}) = 0 \tag{69}$$

Thus the $(T_{k,k}, p_{k,k})$ are roots to Eqs. (68) and (69) and can be determined by straightforward application of Newton–Raphson iteration for finding roots to two simultaneous nonlinear equations. In this application, the technique is best illustrated by working with the truncated Taylor's series approximation to $N_k(T_{k,k}, p_{k,k})$ and $\tau_k(T_{k,k}, p_{k,k})$:

$$N_k^{(m)} - N_k(T_{k,k}^{(j)}, p_{k,k}^{(j)}) \approx \left.\frac{\partial N_k}{\partial T}\right|_{(T_{k,k}, p_{k,k})} (T_{k,k} - T_{k,k}^{(j)}) + \left.\frac{\partial N_k}{\partial p}\right|_{(T_{k,k}, p_{k,k})} (p_{k,k} - p_{k,k}^{(j)}) \tag{70}$$

$$\tau_k^{(m)} - \tau_k^{(m)}(T_{k,k}^{(j)}, p_{k,k}^{(j)}) \approx \left.\frac{\partial \tau_k}{\partial T}\right|_{(T_{k,k}, p_{k,k})} (T_{k,k} - T_{k,k}^{(j)}) + \left.\frac{\partial \tau_k}{\partial p}\right|_{(T_{k,k}, p_{k,k})} (p_{k,k} - p_{k,k}^{(j)}) \tag{71}$$

in which the $N_k(T^{(j)}_{k,k}, p^{(j)}_{k,k})$ and $\tau_k(T^{(j)}_{k,k}, p^{(j)}_{k,k})$ are the band model evaluations for N and τ with values $(T^{(j)}_{k,k}, p^{(j)}_{k,k})$ and the $\partial N_k/\partial T$, $\partial N_k/\partial p$, $\partial \tau_k/\partial T$, $\partial \tau_k/\partial p$ are the partial derivatives of the radiance and transmittance with respect to temperature and pressure, respectively, similarly evaluated at $(T^{(j)}_{k,k}, p^{(j)}_{k,k})$. The only unknown in Eqs. (70) and (71) are the $(T^{(j)}_{k,k}, p^{(j)}_{k,k})$, for which solution is immediate. These values become the new values for the next iterative application of Eqs. (70) and (71). Iteration is continued until the calculated temperature and pressure increments $\Delta T = T_{k,k} - T^{(j)}_{k,k}$, $\Delta p = p_{k,k} - p^{(j)}_{k,k}$ become less than some predetermined value or when the $T^{(j)}_{k,k}$ and $p^{(j)}_{k,k}$ approximation reproduces the experimental values of radiance and transmittance within some predetermined error.

The derivative functions $\partial N_k/\partial T$, $\partial N_k/\partial p$, $\partial \tau_k/\partial T$, and $\partial \tau_k/\partial p$ can become extremely complex to evaluate analytically, depending on which curve of growth is actually used for evaluation. Indeed, analytic evaluation of the derivative function can be the most expensive aspect of the iteration, especially if the plume is modeled by many zones. Finite difference approximation of the partial derivatives is generally more economical and adequate for iteration purposes.

An initial approximation for Eqs. (70) and (71) is necessary. Because the band model equations are nonlinear and coupled, an initial approximation not too far removed from the true value is desired. The solution for the immediately previous zone is an obvious choice for this.

Successive under and over relaxation is an important ingredient to the iterative solution of Eqs. (70) and (71). This is because, depending on wavenumber bandpass, the spatial profiles of temperature and pressure, and the linearized approximation to the band model equations, the process can become quite unstable. Consequently, the ΔT and Δp calculated from Eqs. (70) and (71) are never used in their entirety. Rather, a fraction of the increment is used, the fraction to be used adjusted on each iteration. If the signs of the ΔT and Δp increments remain the same from one iteration to the next, the fraction is increased. When the sign of ΔT or Δp changes, a possible root crossing is indicated and the proportionate amount of the increment is immediately reduced. Continued sign changes continue to reduce the proportion of the ΔT, Δp used to determine the $T_{k,k}$ and $p_{k,k}$.

3.2 Iterative Abel Inversion

An alternative approach to performing the inversion of the radiative transfer equation is one used by Young (1981). This iterative Abel approach similarly uses the band model formulation but uses an Abel transformation to express the measured radiance and transmittance in terms of the cylindrical coordinates of the plume. Since the Abel inversion is strictly correct only for optically thin sources, an iteration technique is required. The general approach is similar to that shown above and for most iterative methods; the calculation of radiance and transmittance (or equivalent width) is compared to the measured values, and the differences are used to calculate corrections to the trial property profiles. Convergence is declared when two successive profiles do not change by more than some preset value. Young similarly used data smoothing to condition the raw data used in the inversion. An advantage of Young's approach is that the entire profile is treated in the inversion and is the result of each iteration step. The data smoothing technique can be incorporated directly into the inversion code. A disadvantage concomitant with this advantage is that if there is a data error, the entire profile information is lost, whereas the onion peel approach described earlier will provide properties to the point at which the data are in error.

In the iterative Abel approach, the average radiance and average equivalent width are calculated from

$$N(z) = 2\int_z^{r_p} N^0(r)\,\frac{S}{d}(r)\,\tau(z)y(z,r)\,\frac{r\,dr}{\sqrt{r^2 - z^2}} \tag{72}$$

$$\frac{\overline{W}}{d}(z) = 2\int_z^{r_p} \frac{S}{d}(r)y(z,r)\,\frac{r\,dr}{\sqrt{r^2 - z^2}} \tag{73}$$

which result from a transformation of the radiative transfer problem with the Lindquist–Simmons approximation [Eqs. (54) and (55)] to cylindrical coordinates. The transformation assumes cylindrical symmetry, with r_p being the maximum radius of the gas volume. Comparison to the original measured profile yields the corrections:

$$\Delta N(z) = N^{(m)}(z) - N(z) = 2\int_z^{r_p} \Delta(Jw) \frac{r\,dr}{\sqrt{r^2 - z^2}} \tag{74}$$

$$\Delta \frac{\overline{W}}{d}(z) = \frac{\overline{W}}{d}^{(m)}(z) - \frac{\overline{W}}{d}(z) = 2\int_z^{r_p} \Delta\left(\frac{S}{d}y\right) \frac{r\,dr}{\sqrt{r^2 - z^2}} \tag{75}$$

in which the source function $J = N^0 S/d$ and transmittance derivative function $w = \tau y$ are used to compress the notation. The corrections to the kernels $\Delta(Jw)$ and $\Delta(S/d)y$ are found by performing the Abel inversion. The correction yields new $N(z)$, $(\overline{W}/d)(z)$, and the process is iterated. When the iteration has converged, the temperature profile is determined from

$$T(r) = \frac{C_2\bar{\nu}}{\ln\{[C_1\bar{\nu}^3(S/d)(r)/\pi J(r)] + 1\}} \tag{76}$$

in which C_1 and C_2 are the constants from the Kirchoff-Planck law, $C_1 = 3.741 \times 10^{-16}$ W m^2 and $C_2 = 1.4389$ cm K. With $T(r)$ known, the $(S/d)(r)$ is known, and pressure is found from

$$p(r) = \frac{(S/d)(r)}{\overline{K}_0(r)} \tag{77}$$

3.3 Errors Propagation

It is important to determine the effects of the usually present experimental measurement errors on the final temperature and partial pressure determinations. Because of the nonlinear coupling between Eqs. (61)–(62), and the Curtis–Godson approximation, the errors near the center of the plume are correlated to those to the outer portions of the plume, even though the measurements themselves may not be. Determination of resultant uncertainties must then include this coupling and correlation. Limbaugh and Kneile (1984) show that a method to accomplish this can be approached through a matrix formulation for variance–covariance propagation. The iteration equations represented by Eqs. (70) and (71) can be collected into a partitioned matrix form:

$$\left[\begin{array}{c|c} \dfrac{\partial N_k}{\partial T} & \dfrac{\partial N_k}{\partial p} \\ \hline \dfrac{\partial \tau_k}{\partial T} & \dfrac{\partial \tau_k}{\partial p} \end{array}\right] \left[\begin{array}{c} \Delta T \\ \hline \Delta p \end{array}\right] = \left[\begin{array}{c} \Delta N \\ \hline \Delta \tau \end{array}\right] \tag{78}$$

where

$$\left[\frac{\partial N_k}{\partial T}\right] = \begin{bmatrix} \dfrac{\partial N_1}{\partial T_1} & 0 & 0 & \cdots \\ \dfrac{\partial N_2}{\partial T_1} & \dfrac{\partial N_2}{\partial T_2} & 0 & \cdots \\ \dfrac{\partial N_3}{\partial T_1} & \dfrac{\partial N_3}{\partial T_2} & \dfrac{\partial N_3}{\partial T_3} & \cdots \\ \vdots & \vdots & \vdots & \end{bmatrix} \tag{79}$$

$$\left[\frac{\partial \tau_k}{\partial p}\right] = \begin{bmatrix} \dfrac{\partial \tau_1}{\partial p_1} & 0 & 0 & \cdots \\ \dfrac{\partial \tau_2}{\partial p_1} & \dfrac{\partial \tau_2}{\partial p_2} & 0 & \cdots \\ \dfrac{\partial \tau_3}{\partial p_1} & \dfrac{\partial \tau_3}{\partial p_2} & \dfrac{\partial \tau_3}{\partial p_3} & \cdots \\ \vdots & \vdots & \vdots & \end{bmatrix} \tag{80}$$

and the submatrices of the other derivatives have a similar triangular form.

Identifying the matrices

$$A = \left[\begin{array}{c|c} \dfrac{\partial N_k}{\partial T} & \dfrac{\partial N_k}{\partial p} \\ \hline \dfrac{\partial \tau_k}{\partial \tau} & \dfrac{\partial \tau_k}{\partial p} \end{array}\right], \tag{81}$$

$$C = \left[\begin{array}{c} \Delta T \\ \hline \Delta p \end{array}\right], \tag{82}$$

$$M = \left[\begin{array}{c} \Delta N \\ \hline \Delta \tau \end{array}\right], \tag{83}$$

Eq. (78) has an immediate solution,

$$C = A^{-1}M \tag{84}$$

Limbaugh and Kneile (1984) show that the variance–covariance matrix for the calculated temperature and pressure can be expressed as

$$V_c = A^{-1}V_m(A^{-1})^T \tag{85}$$

where V_M is the variance–covariance matrix for the radiance and transmittance data used in the inversion. The diagonal elements of V_M are the variance (or squares of the standard deviations), and the off-diagonal elements are the covariances.

The experimental measurements are usually independent and uncorrelated so that the variance–covariance matrix representing the raw data is usually diagonal, with the off-diagonal elements being zero and the diagonal elements being the square of the observed standard deviation. If the raw data are smoothed in some fashion, then this is no longer the case and the off-diagonal elements, or covariances, will not be zero. However, the variance–covariance matrix for the smoothed data can be obtained by casting the curve fit equations into the form of linear equations and applying Eq. (85). Shelby (1976) (and Shelby and Limbaugh, 1977) illustrates the development of the relevant matrix formulation for a least squares spline fit procedure as applied to Abel inversion of plasma diagnostics data.

4 EXAMPLE APPLICATIONS

Two studies showing validity of the coupled band model formulation in an onion peel radial inversion are described by Brewer and Limbaugh (1972) and more extensively by Brewer (1975). In that work laboratory measurements were made on a flat flame burner using hydrogen-air and methane-air for fuel. The flat flame burner premixed the fuel and air and was designed such that the reactant gases were distributed over the face of the burner with uniform velocity. This results in a uniform temperature and reactant species concentration across the face of the burner. There were no guard gases around the periphery of the burner to prevent mixing of atmospheric gases at the outer edge, so that dilution and deviation of results from established theoretical flat flame burner performance near the outer edge was to be expected.

Results of the hydrogen-air burner measurements and subsequent radial inversion are shown in Figs. 10 and 11. Measurements of radiance and transmittance were obtained approximately 1/2 cm above the grid with the burner operating at stoichiometric

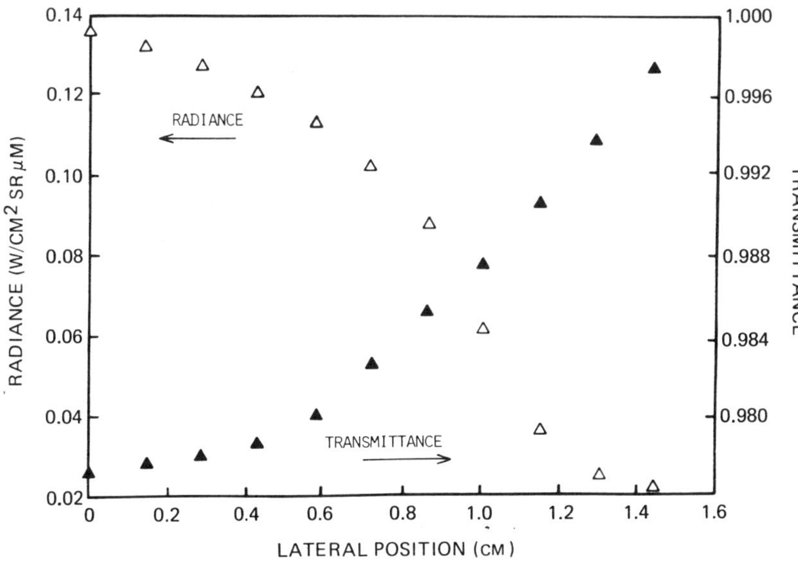

Fig. 10 Radiance and transmittance from hydrogen–air flame, $\bar{\nu} = 4000\ cm^{-1}$ (from Brewer, 1975).

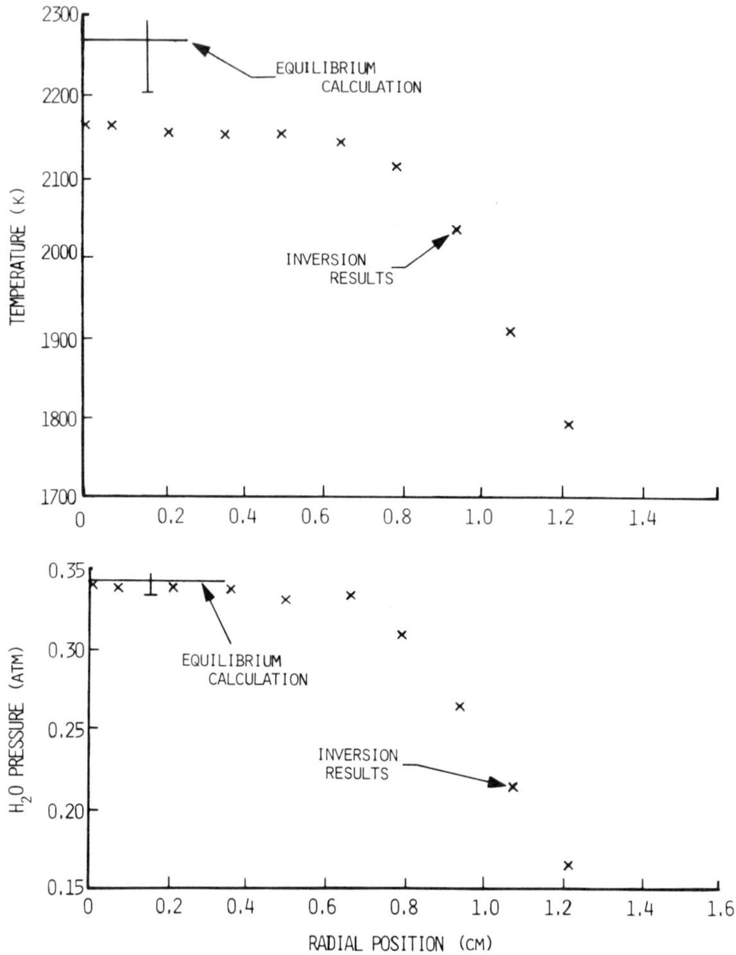

Fig. 11 Temperature and H_2O partial pressure profiles from hydrogen-air flame, $\bar{\nu}$ = 4000 cm^{-1} (from Brewer, 1975).

conditions. The resultant shape of the profiles confirms the essential uniformity of the profiles of the burner to a radius of about 0.8 cm. The influence of mixing of cold external gases over the last 0.5 cm of the flame is obvious. Measurements were made at 2.5 μm in the 2.7-μm band of H_2O. No thermocouple profiles were taken in the experiment for independent direct confirmation of results; however, an equilibrium calculation showed agreement within 5% of the centerline results from the radial inversion. The random errors propagation estimates were not available at the time of that work. The uncertainty bar is an estimate of the errors in the calculation due to the uncertainty in the burner operating parameters.

Results of methane-air flat flame burner experiments are shown in Figs. 12 and 13. In these experiments measurements were made in both the 2.7-μm band of H_2O and the 4.3-μm band of CO_2. It is seen that the agreement between the temperature profiles obtained independently in both H_2O and CO_2 bands is excellent, except near the outer edge where there is mixing of am-

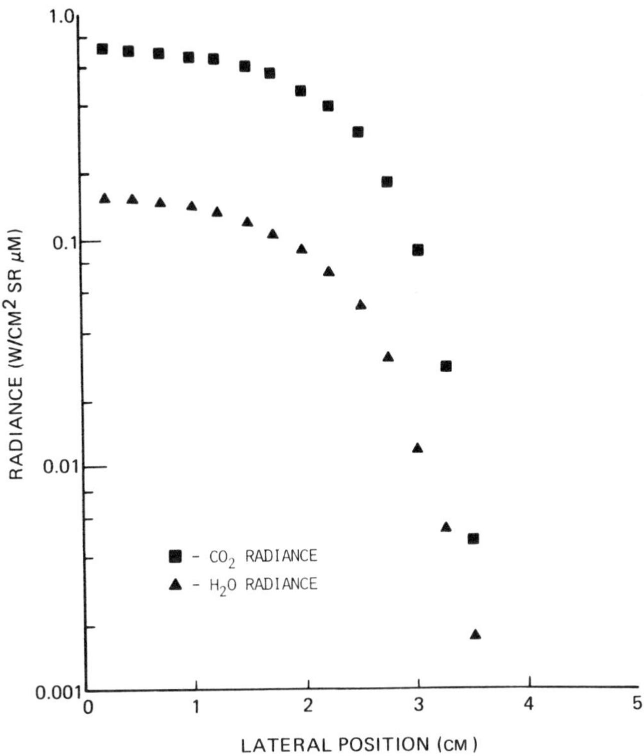

Fig. 12a Radiance data from methane–air flame, $\bar{\nu}_{H_2O} = 4000$ cm^{-1}, $\bar{\nu}_{CO_2} = 2222$ cm^{-1} (from Brewer, 1975).

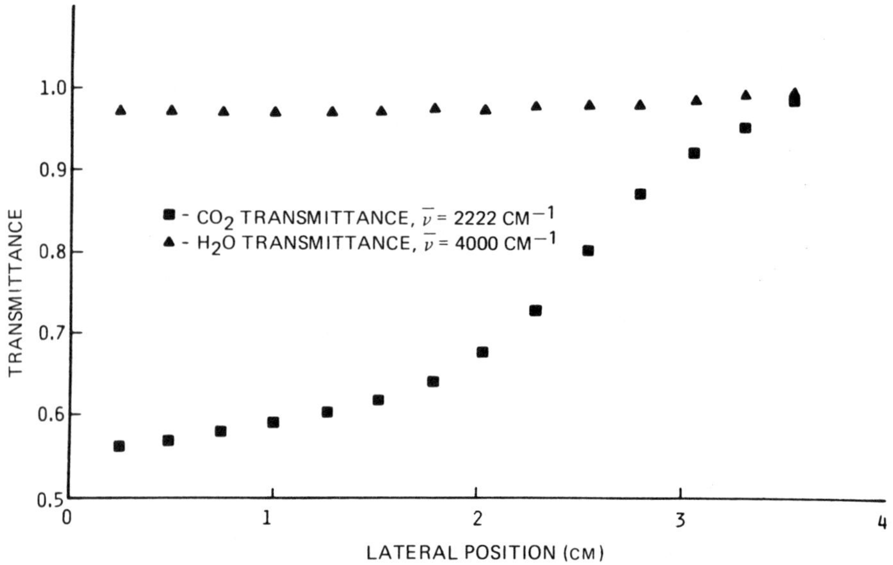

Fig. 12b Transmittance data from methane–air flame, $\bar{\nu}_{H_2O}$ = 4000 cm^{-1}, $\bar{\nu}_{CO_2}$ = 2222 cm^{-1} (from Brewer, 1975).

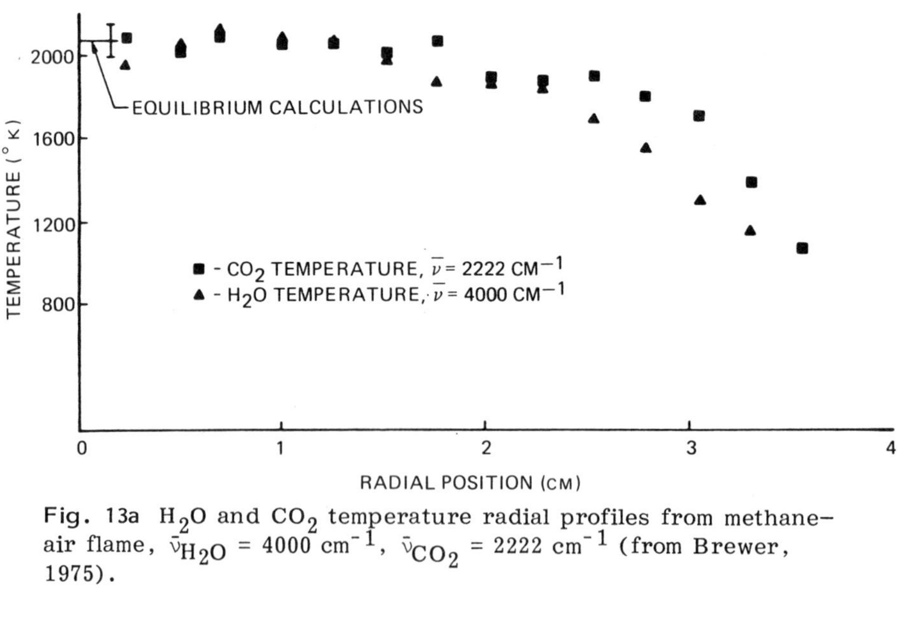

Fig. 13a H_2O and CO_2 temperature radial profiles from methane–air flame, $\bar{\nu}_{H_2O}$ = 4000 cm^{-1}, $\bar{\nu}_{CO_2}$ = 2222 cm^{-1} (from Brewer, 1975).

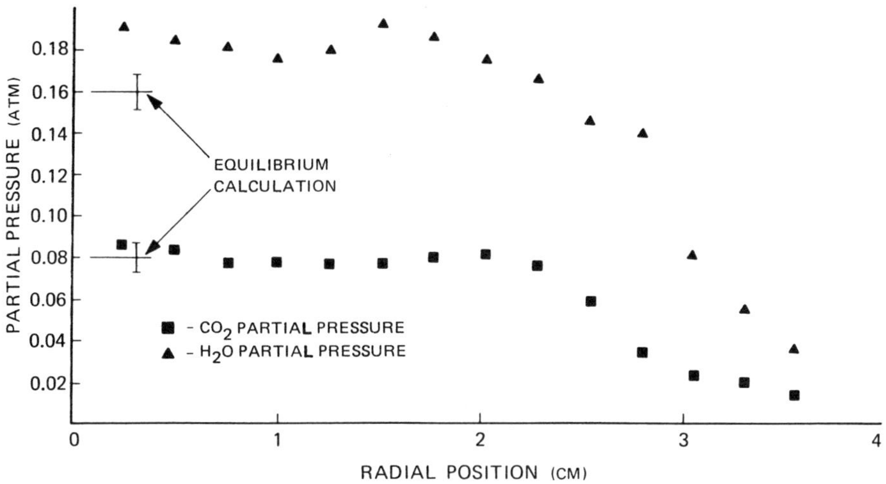

Fig. 13b H_2O and CO_2 partial pressure profiles from methane–air flame, $\bar{\nu}_{H_2O} = 4000\ cm^{-1}$, $\bar{\nu}_{CO_2} = 2222\ cm^{-1}$ (from Brewer, 1975).

bient atmospheric gases and the transmittance of water vapor has become near 1.0. Similarly, agreement with an equilibrium calculation at the centerline is excellent. The error bar is an estimated uncertainty imposed on the equilibrium calculation because of the known nonideal conditions of the experiment.

The partial pressure profiles of H_2O and CO_2 are shown in Fig. 13 compared to the calculated equilibrium conditions. Here the agreement of CO_2 with equilibrium is again excellent, while the H_2O is about 12% higher than calculation. This discrepancy is attributed to the fact that the air used for those measurements was not dried and the relative humidity of the air can easily cause this difference.

Brewer (1975) similarly reported results of measurements made on a full-scale turbojet engine in an outdoor test stand. A schematic of the engine and the E/A scanning arrangement are shown in Fig. 14. Measurements were made at 2.5 μm in the 2.7-μm water vapor band. Points from the smoothed radiance and transmittance profiles are shown in Fig. 15. The resulting temperature and water vapor pressure profiles are shown in Figs. 16a and 16b. At the time of that work, the errors propagation analysis was not available, and there is no estimate of the propagated uncertainties. Shown in Figs. 16a and 16b is the result of an equilibrium

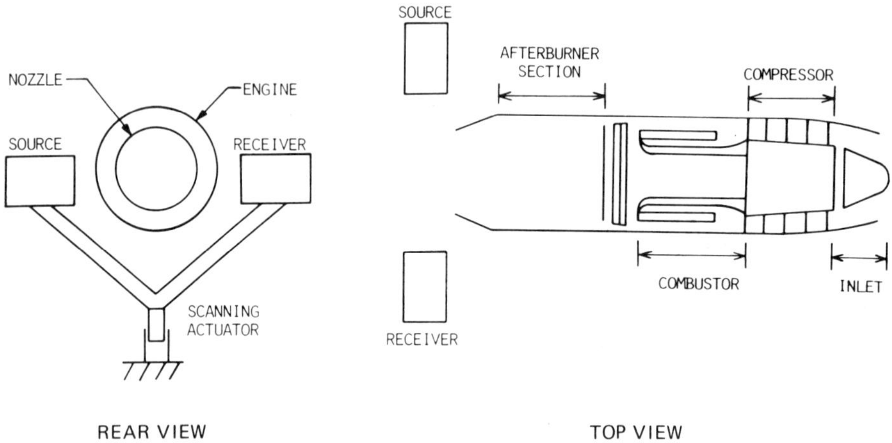

Fig. 14 Schematic of jet engine and experimental arrangement for exhaust radiance and transmittance measurements (adapted from Brewer, 1975).

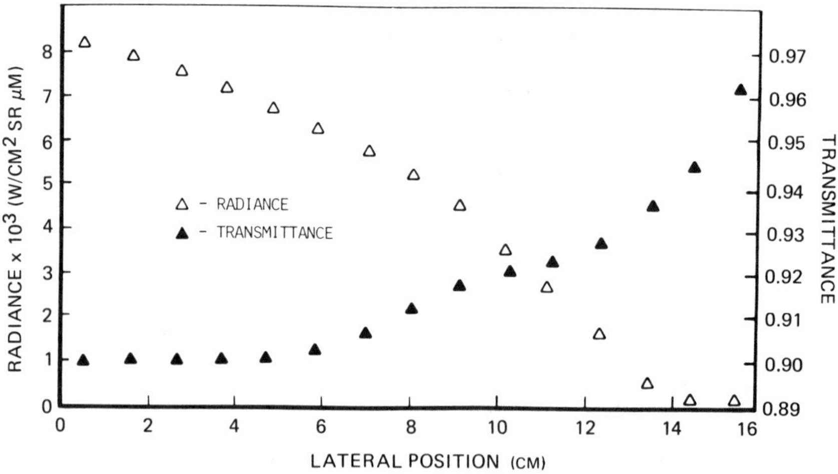

Fig. 15 Radiance and transmittance data from jet engine exhaust, $\bar{\nu}$ = 4000 cm^{-1} (adapted from Brewer, 1975).

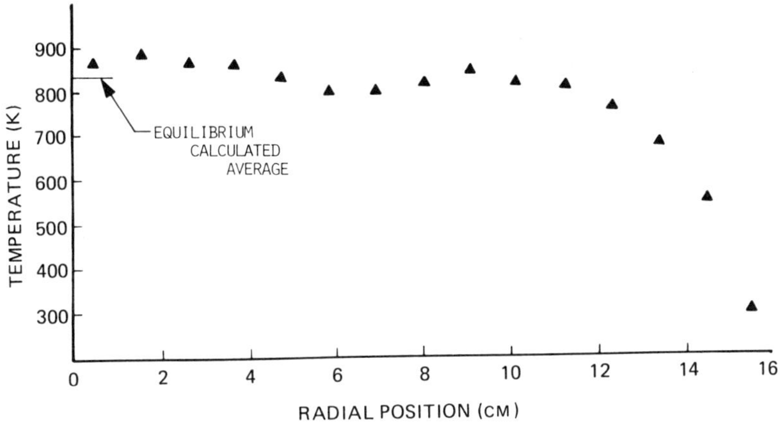

Fig. 16a H_2O Temperature radial profile from jet engine exhaust, $\bar{\nu}$ = 4000 cm^{-1} (from Brewer, 1975).

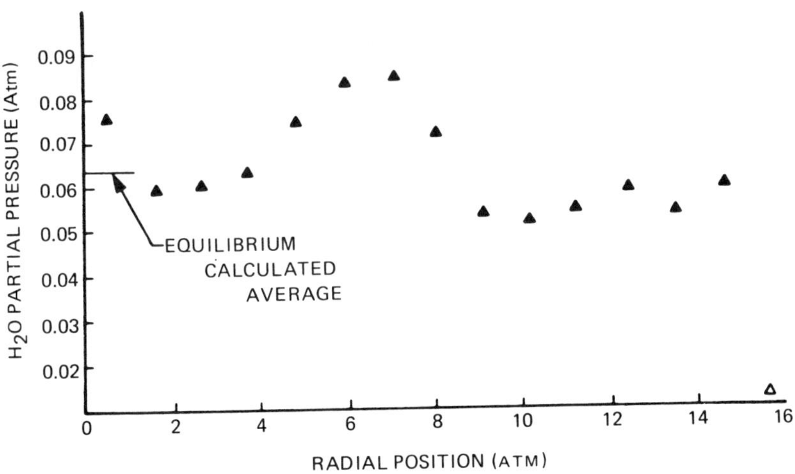

Fig. 16b H_2O partial pressure radial profile from jet engine exhaust, $\bar{\nu}$ = 4000 cm^{-1} (from Brewer, 1975).

chemistry calculation which was believed to give a good representation of the centerline value for the engine. The observed structure in the partial pressure profile was unexpected and unexplained, and repeat measurements were not made. At the time it was believed that this may have been due to some unknown data error; however, unreported calculations based on probe measurements have since shown the structure is to be expected.

The results of applying the onion peel inversion and error propagation analysis to the example smoothed radiance and transmittance profiles of Fig. 8 have been reported by Limbaugh and Kneile (1984) and are shown in Fig. 17. A step function format is used to emphasize the geometry model used in the inversion. The propagated random uncertainties determined by the procedure described above (and in more detail by Limbaugh and Kneile, 1984) are included as the $\pm 2\sigma$ bands. The initial $\pm 2\sigma$ uncertainty for the raw data was indicated on Fig. 8. The inversion results shown in Fig. 17, compared to the data of Fig. 8, illustrate some rather general characteristics of such E/A inversions. The largest uncertainties generally appear near the outer edge where the signal-to-noise ratio is small. Propagated uncertainties are also generally larger near the centerline where the information in the data is least. Indeed, the best errors propagation characteristics occur where the data are changing. This is because, as the inversion proceeds from the outer edge inward, the change in the (N, τ) data is in large measure due to the contribution of the inner unknown zone, and, for an appreciable change in data, the significance of the inner zone is appreciable. For nearly flat profiles, such as in Fig. 8, the contribution of the inner zone is of correspondingly little significance, and large uncertainties can result. Higher quality data are required in this case. As a general rule, monotonic data (decreasing for radiance and increasing for transmittance) from the centerline to the outer edge will give inversion with the smallest propagated uncertainties.

The E/A technique does have the ability to resolve spatial structure, e.g., the strong pressure gradient at $r/r_e = 0.65$ in Fig. 17. In this case, the transmittance (Fig. 8) goes through a minimum at about $r/r_e = 0.65$, while there is no corresponding strong change in the radiance profile. Consequently, the partial pressure shows a strong radial pressure gradient, while the temperature shows a smaller, though still apparent, change. Data for inversion from regions in which radial gradients in the gas properties were known to exist should be of high quality.

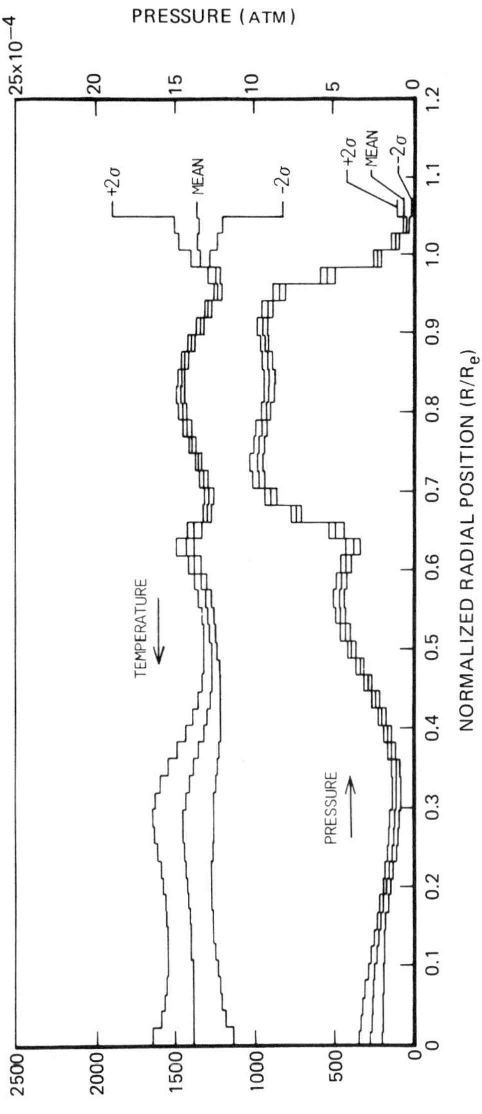

Fig. 17 CO_2 temperature and partial pressure radial profiles from rocket exhaust data of Fig. 8. (Reprinted with permission from the *Journal of Quantitative Spectroscopy and Radiative Transfer*, *31*:161–171, Limbaugh and Kneile, "Uncertainties Propagation for Combustion Diagnostics Using Infrared Band Models," Copyright 1984, Pergamon Press, Ltd.)

5 CONCLUDING REMARKS

The E/A techniques described in the foregoing discussion were shown, in the illustrated applications, to allow the determination of spatial profiles of the gas temperature and partial pressure. The success of the technique is highly dependent on integrating measurements and analysis technology. The analytical model used must, as nearly as possible, represent the physical conditions at which the data are acquired. For example, the development here has been for pure gases. If particles are present, their contribution to the data must be included in the model for the measured radiance and transmittance. Instrumentation bandpasses must be used which provide data from only one molecule, or a modification of the model must be made to include contribution from all the molecules within the selected bandpass. The present developments are for a cylindrical geometry, but there is no difficulty in principle for other geometries. As needs arise and the measurement capabilities and analytical models improve, the range of applications should correspondingly broaden.

ACKNOWLEDGMENT

The author wishes to acknowledge and express appreciation to W. K. McGregor for his contributions through continued discussions and to L. E. Brewer for his contributions in the early stages of development of the E/A technique.

REFERENCES

Armstrong, B. H. (1968). Analysis of the Curtis-Godson approximation and radiation transmission through inhomogeneous atmosphere. *J. Atm. Sci.*, *25*: 312-322.

Balasubramanian, R., Norrie, D. H., and DeVries, G. (1979). The application of the least squares finite element method to Abel's integral equation. *Int. J. Numer. Meth. Eng.*, *14*: 201-209.

Barr, W. L. (1962). Method for computing the radial distribution of emitters in a cylindrical source. *J. Opt. Soc. Am.*, *52*: 855-888.

Bernstein, L. S. (1980). Band model parameters for the parallel bands of linear triatomic molecules—I. Theory. *J. Quant. Spectrosc. Radiat. Trans.*, *23*: 157–167.

Bernstein, L. S., Robertson, D. C., and Conant, J. A. (1980). Band model parameters for the 4.3 μm band from 200 to 3000 K—II. Prediction, comparison to experiment, and application to plume emission-absorption calculations. *J. Quant. Spectrosco. Radiat. Trans.*, *23*: 169–185.

Blair, D. W. (1974). An analysis of error propagation in Abel inversion of spectral emission-absorption data. *J. Quant. Spectrosc. Radiat. Trans.*, *14*: 325–337.

Brewer, L. E. and Limbaugh, C. C. (1972). Infrared band model technique for combustion diagnostics. *Appl. Opt.*, *11*: 1200–1204.

Brewer, L. E. (1975). "Development and Application of an Optical Technique for Determining Temperature and Species Concentrations in Radiating Combustion Gases." Ph.D. Thesis, Von Karman Institute for Fluid Dynamics, Brussels, Belgium.

Cremers, C. J. and Birkebak, R. C. (1966). Application of the Abel integral equation to spectrographic data. *Appl. Opt.*, *5*: 1057–1064.

Curtis, A. R. (1952). A statistical model for water vapor absorption. *Quart. J. Roy. Meteorol. Soc.*, *78*: 638–640.

Dahm, W. K. (1967). Introduction to the Problem of Rocket Base Heating and to the Behavior of Liquid Propellant Rocket Plumes. *Specialist Conference on Molecular Radiation and Its Application to Diagnostic Techniques*, NASA-TMX-53711, Marshall Space Flight Center, Huntsville, Alabama, Oct. 5–6, 1967.

Deutsch, M. and Beniaminy, I. (1983). Inversion of Abel's integral equation for experimental data. *J. Appl. Phys.*, *54*: 137–143.

Elsasser, W. M. (1942). Heat transfer by infrared radiation in the atmosphere. *Harvard Meteorological Studies No. 6.* Harvard University Press, Cambridge, Massachusetts.

Ferriso, C. C., Ludwig, C. B. and Thomson, J. A. L. (1966). Empirically determined infrared absorption coefficients from 300–3000 K. *J. Quant. Spectrosc. Radiat. Trans.*, *6*: 241–273.

Freeman, M. P. and Katz, S. (1960). Determination of the radial distribution of brightness in a cylindrical luminous medium with self absorption. *J. Opt. Soc. Am.*, *50*: 826–830.

Glasser, J., Chapelle, J. and Boettner, J. C. (1978). Abel inversion applied to plasma spectroscopy: A new interactive model. *Appl. Opt.*, *17*: 3750–3754.

Goody, R. M. (1964). *Atmospheric Radiation I. Theoretical Basis*. Oxford Univ. Press, London and New York.

Goody, R. M. (1964b). The transmission of radiation through an inhomogeneous atmosphere. *J. Atmo. Sci.*, *21*: 575–581.

Godson, W. L. (1953). The evaluation of infrared-radiative fluxes due to atmospheric water vapor. *Quart. J. Roy. Meteorol. Soc.*, *79*: 367–379.

Herget, W. (1967). Temperature and Concentration Measurements in Model Exhaust Plumes Using Inversion Techniques. *Specialist Conference on Molecular Radiation and Its Application to Diagnostic Techniques*. NASA-TMX-53711, Marshall Space Flight Center, Huntsville, Alabama, Oct. 5–6, 1967.

Hommert, P. J., Viskanta, R. and Mellor, A. M. (1977). Flame temperature measurements by spectral remote sensing. *Combustion and Flame*, *30*: 295–308.

Hughey, B. J. and Santavicca, D. A. (1982). A comparison of techniques for reconstructing axisymmetric reacting flows from absorption measurements. *Combustion Sci. Technol.*, *29*: 167–190.

Kaplan, L. D. and Eggers, D. F. (1956). Intensity and line width of the 15-micron CO_2 band, determined by a curve-of-growth method. *J. Chem. Phys.*, *25*: 876.

Kaplan, L. D. (1959). A method for calculation of infrared flux for use in numerical models of atmospheric motion. *The Atmosphere and the Sea in Motion*, Rockefeller Institute Press and Oxford University Press, New York, pp. 170–177.

Koo, J., Goulard, R., Ludwig, C. B. and Malkmus, W. (1979). "Radiative properties of combustion gases," Paper 79-1041, AIAA 14th Thermophysics Conference, Orlando, Florida.

Krakow, B. (1966). Spectroscopic temperature profile measurements in inhomogeneous hot gases. *Appl. Opt.*, *5*: 201–209.

Krakow, B., Babrov, H. J., Maclay, G. J. and Shabott, A. L. (1966). Use of the Curtis–Godson approximation in calculations of radiant heating by inhomogeneous hot gases. *Appl. Opt.*, *5*: 1791–1801.

Krakow, B., Kiech, E. L. and McAdoo, H. A. (1977). *Isolated CO lines for use in combustion gas diagnostics.* AEDC-TR-77-11, Arnold Engineering Development Center, Arnold Air Force Station, Tennessee. Available from DTIC, AD-A044053.

Kunitomo, T. and Osumi, M. (1975). Narrow band model parameters for the CO_2 4.3-μm band. *J. Quant. Spectrosc. Radiat. Trans.*, *15*: 345–356.

Ladenburg, R. and Reiche, F. (1913). Uber selektive absorption, *An. Phys.*, *42*:

Limbaugh, C. C. (1977). Excited state distribution function of argon in a freely expanding arcjet plume, *Progress in Astronautics and Aeronautics*, New York, pp. 933–948.

Limbaugh, C. C. and Kneile, K. R. (1984). Uncertainties propagation for combustion diagnostics using infrared band models, *J. Quant. Spectrosc. Radiat. Trans.*, *31*: 161–171.

Lindquist, G. G. and Simmons, F. S. (1972). A band model formulation for very nonuniform paths. *J. Quant. Spectrosc. Radiat. Trans.*, *12*: 807–820.

Lindquist, G. G., Arnold, C. B. and Spellicy, R. L. (1975). *Atmospheric Absorption Applied to Plume Emission: Experimental and Analytical Investigations of Hot Gas Emission Attenuated by Cold Gas.* ERIM Report No. 102700-20-F, Environmental Research Institute of Michigan, Ann Arbor, Michigan.

Ludwig, C. B., Malkmus, W., Reardon, J. E. and Thomson, J. A. L. (1973). *Handbook of Infrared Radiation from Combustion Gases.* (R. Goulard and J. A. L. Thomson, eds.) NASA SP-3080, National Aeronautics and Space Administration, Scientific and Technical Information Office, Washington, D.C.

McGregor, W. K. and Brewer, L. E. (1966). Equivalence of electron and excitation temperatures in argon plasmas. *The Physics of Fluids*, *9*: 826.

Pearce, W. J. (1958). Calculation of the Radial Distribution of Photon Emitters in Symmetric Sources. *Conference on Extremely High Temperatures.* John Wiley and Sons, Inc., New York, pp. 123–135.

Penner, S. S. (1959). *Quantitative Molecular Spectroscopy and Gas Emissivities*, Addison-Wesley Publishing Co., Inc., Reading, Massachusetts.

Ray, S. and Semerjian, H. G. (1983). "Laser Tomography for Simultaneous Concentration and Temperature Measurements in Reacting Flows." Paper AIAA-83-1553, AIAA 18th Thermophysics Conference, Montreal, Canada.

Rodgers, C. D. and Williams, A. P. (1974). Integrated Absorptance of a Spectral Line with the Voigt Profile, *J. Quant. Spectrosc. Radiat. Trans.*, *14*: 319–373.

Shelby, R. T. (1976). *Abel Inversion Error Propagation Analysis*, Masters Thesis, University of Tennessee Space Institute, Tullahoma, Tennessee.

Shelby, R. T. and Limbaugh, C. C. (1977). *Smoothing Technique and Variance Propagation for Abel Inversion of Scattered Data*, AEDC-TR-76-163, Arnold Engineering Development Center, Arnold Air Force Station, Tennessee.

Simmons, F. S., Yamada, H. Y. and Arnold, C. G. (1969). *Measurement of Temperature Profiles in Hot Gases by Emission-Absorption Spectroscopy*, NASA-CR-72491, NASA Lewis Research Center, Cleveland, Ohio. Available from NTIS, N69-32557.

Tourin, R. H. (1966). *Spectroscopic Gas Temperature Measurement*. Elsevier Publishing Company, New York.

Walshaw, C. D. and Rodgers, C. D. (1963). The effect of the Curtis–Godson approximation on the accuracy of radiative heating rate calculations. *Quart. J. Roy. Meteorol. Soc.*, *89*: 122–130.

Wang, J. Y. (1976). Laser absorption methods for simultaneous determination of temperature and species concentrations through a cross section of a radiating flow. *Appl. Opt.*, *15*: 768–773.

Wang, J. Y. (1977). Passive optical diagnostics of combustion flows. *Appl. Opt.*, *16*: 627–634.

Young, S. J. (1975). Band model formulation for inhomogeneous optical paths. *J. Quant. Spectrosc. Radiat. Trans.*, *15*: 483–501.

Young, S. J. (1975a). Addendum to: Band model formulation for inhomogeneous optical paths. *J. Quant. Spectrosc. Radiat. Trans.*, *15*: 1137–1140.

Young, S. J. (1976). *Band Model Parameters for the 2.7-μm Bands of H_2O and CO_2 in the 100 to 3000 K Temperature Range*, SAMSO-TR-209, Space amd Missile Systems Organization, Air Force Systems Command, Los Angeles, California.

Young, S. J. (1976a). *Band Model Parameters for the 4.3-μm Fundamental Band of CO_2 in the 100–3000 K Temperature Range*, SAMSO-TR-76-35, Space and Missile Systems Organization, Air Force Systems Command, Los Angeles, California.

Young, S. J. (1977). Nonisothermal band model theory. *J. Quant. Spectrosc. Radiat. Trans.*, *18*: 1–28.

Young, S. J. (1977a). Evaluation of nonisothermal band models for H_2O. *J. Quant. Spectrosc. Radiat. Trans.*, *18*: 29–45.

Young, S. J. (1978). Curve-of-growth function for a random array of Voigt lines. *J. Quant. Spectrosc. Radiat. Trans.*, *20*: 331–332.

Young, S. J. (1979). Analysis and computation of the derivative function for an isolated Lorentz line. *J. Quant. Spectrosc. Radiat. Trans.*, *22*: 187–194.

Young, S. J. (1981). Iterative Abel inversion of optically thick, cylindrical symmetric radiation sources. *J. Quant. Spectrosc. Radiat. Trans.*, *25*: 479–481.

6

Nondispersive Infrared Absorption Monitors for Trace Gases

DANIEL I. SEBACHER *NASA Langley Research Center, Hampton, Virginia*

1 INFRARED MEASUREMENT TECHNIQUES

All infrared (IR) analyzers require a source, a detector, an absorbing medium, and a method of wavelength selection. Chapters 1–5 have chiefly treated resolved emission spectra observed by Fourier transform infrared spectrometers. Chapter 3 also gave samples of FT-IR absorption spectra. For both methods, there will be applications which do not require as detailed spectral information, and for which lower resolution devices are adequate. Lower resolution devices can offer ease of operation and simplicity in a tradeoff against resolution. These devices include grating and prism spectrometers and bandpass filters. For absorption applications, this chapter describes a similarly simple technique which still provides the selectivity associated with high spectral resolution.

The nondispersive infrared technique does not use prisms or gratings but uses a sample of gas to provide a selective filter for radiation absorbed in a gas mixture containing the specified gas. When two similar cells, one containing the specifying gas and one containing a nonabsorbing gas or a vacuum, are placed alternately in a light beam, the radiation measured by the detector collecting the beam is differentially modulated only at the wavelengths at which the specifying gas absorbs. If the optical depth of the specifying gas cell is increased to where the radiation is very strongly absorbed at these characteristic wavelengths, then any small additional absorption of the beam by the specified gas located outside the cells will cause a small decrease in the modulated signal, since the additional absorption will produce a larger signal change when the nonabsorbing cell is in the beam than when the specifying cell is in the beam. The difference in signal change is a result of the spectral nature of molecular absorption of light. For a given concentration of specified gas outside the cell, the largest signal change would occur if the optical depth of the specifying cell could be increased to where no additional outside absorption is detected when the specifying cell is in the beam. In this manner, the specifying cell acts as a selective gas filter for radiation absorbed by a mixture of atmospheric constituents.

2 RESOLUTION

Resolution attained with the gas filter correlation technique may be exceptionally high but depends on the spectral line broadening, temperature, and optical depth of the specifying gas selected. The definition of the resolving power or resolution of a dispersive instrument is given by

$$R = \frac{\bar{\nu}}{\Delta \bar{\nu}} \tag{1}$$

where $\Delta \bar{\nu}$ is the smallest difference in wavenumber of identical spectral lines that can be discriminated, and $\bar{\nu}$ is the average of the wavenumbers $\bar{\nu}_1$ and $\bar{\nu}_1 + \Delta \bar{\nu}$. The Rayleigh criterion for resolution was introduced in Chap. 3. In the context of FT-IR instrument line shapes, it was defined as stating that two lines of equal intensity are resolved when the maximum of one falls on the first minimum of the other. In the case of the sinc^2 function, the full width as half maximum is 0.9 of the $\Delta \bar{\nu}$ for the Rayleigh criterion. For the nondispersive gas filter technique, frequency selection results from the absorption by the specifying gas, and the resolution is determined by the width of the rotational lines within the gas absorption band spectra. Only energy at frequencies within these narrow intervals results in a change in the analyzer output.

The Rayleigh criterion as defined above can also be shown to correspond to having the value at the midpoint between the two overlapping resonances equal to about 0.8 of the peak of either isolated resonance. This definition is also applicable to molecular line shapes. For the Lorentzian shape, Eq. (35), it is attained when

$$\Delta \bar{\nu} = 1.2 \times 2\gamma \tag{2}$$

where γ is the line half-width at half maximum. The pressure-broadened linewidths in the rotational bands of interest in this study are in the range of 0.005 to 0.02 cm^{-1}, while the widths produced by the Doppler effect are from 0.001 to 0.01 cm^{-1} (Herzberg, 1950). Since the nondispersive gas filter technique uses a number of absorption lines which differ in half-width and whose values are of questionable accuracy, the actual resolution of this type of instrument is generally not calculated. However, because it is a function of the line half-width, the resolution of a gas filter correlation analyzer should be equal to or better than typical spectrometers, either dispersive instruments whose resolution is determined by slit-width limitations or interferometers which are limited by mirror displacement and stability. Since linewidth is a function of pressure and optical depth, the selectivity can be increased by decreasing the partial pressure and optical length of the specifying gas cell. Gas filter correlation also offers the advantage of using the contribution of a large number of absorption lines of a band system without decreasing resolution; thereby the radiation level is increased above that generally available when using conventional spectroscopic instruments.

We can analyze gas-filter analyzers in terms of the multiplex and throughput advantages already discussed in connection with FT-IR spectrometers in Sec. 2 of Chap. 3. As pointed out by Davies, Barringer, and Dick (1982), a gas-filter correlation spectrometer is in fact fully multiplexed, in that all wavelengths are in phase at the detector and there is no signal power cancellation due to phase shifts for different wavelengths. The use of the zero-retardation signal from an FT-IR spectrometer as described in Chap. 2 shares this advantage. The throughput advantage of nondispersive instruments is also large, since the gas cell which determines the resolution imposes none of the angular restrictions involved in interferometers and dispersive spectrometers and since the gas cells can be made large more easily than other optical components.

3 NONDISPERSIVE INSTRUMENT DESIGN

Although nondispersive infrared techniques have been applied to gas measurements for over three decades (Wright and Herscher, 1946; Fastie and Pfund, 1947; Fowler, 1949; Bartle et al., 1972; Ward and Zwick, 1975; Wiens and Zwick, 1977; Sebacher, 1978), the need for higher sensitivity in air-pollution monitoring of trace gases has accelerated the development of these instruments. (Wiens and Zwick (1977) also review dispersive correlation spectrometers, in which a mask placed at the exit focal plane of a prism grating instrument allows for optical processing of spectra to yield a single measured quantity and ignore interfering gases, but which have the throughput disadvantages of such instruments; and correlation interferometers, which have improved throughput but not the simplicity of gas-filter devices.) Review articles on the evolution of advanced spectroscopic techniques can be found in the literature (Hodgeson et al., 1973; Hanst et al., 1973). Improvements are being accomplished by modification of the optical systems and by advances in the electronic signal processing. Since instruments are needed that can measure concentrations of parts per million over a 1-m path length of light (ppm-m), the electronic processing system must detect and measure a signal difference in the range of one part in a thousand.

The basic elements of an early gas-filter analyzer (Fastie and Pfund, 1947) are shown in Fig. 1. A nonselective IR source passes light through a sampling cell which is sealed with broad band IR filter windows. The detector is a gas-filled cell containing thermocouples to measure the temperature difference between the wall of the cell and the gas in the cell.

If CO_2 is to be measured, the detector is filled with pure CO_2. Radiation from the source heats the detector cell gas, but only in the wavelength in which CO_2 absorbs. Therefore, other absorbing gases in the sampling cell which do not have absorption bands which overlap the CO_2 bands have no effect on the heating of the CO_2 in the detector cell. However, when CO_2 enters the sample cell, it absorbs at the same wavelength which heats the gas in the detector cell and therefore causes a change in the output of the detector-cell thermocouples. This technique makes the analyzer selective to CO_2 radiation.

An early version of the negative-filtering method (Wright and Herscher, 1946) is illustrated in Fig. 2. The light source and sample cell are similar to the first design, but two thermopile detectors are now used with gas-filter cells in front of each detector. The two detectors are connected in series so that no net thermopile output is produced for identical radiation through the cells.

If specifying cell F_1 is now filled with a pure absorbing gas such as CO_2, and reference cell F_2 is filled with a nonabsorbing gas, the detector T_1 will be desensitized to radiant energy at the CO_2 wavelengths because of the heavy filtering of CO_2 in gas-

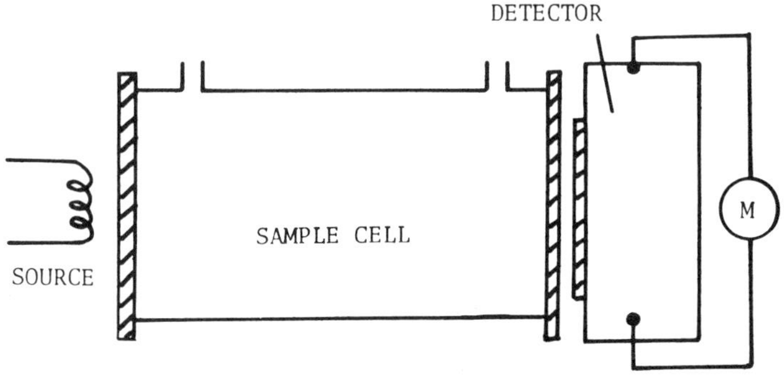

Fig. 1 An early version of the gas-filter analyzer with positive filtering.

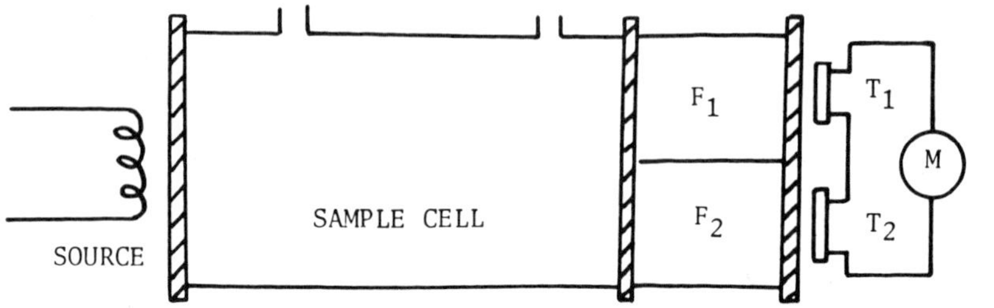

Fig. 2 Early gas-filter analyzer using the negative-filtering method.

filter cell F_1. On the other hand, detector T_2 is still sensitive to radiation at the CO_2 wavelengths and responds to the entry of CO_2 in the sample cell. Both detectors are equally sensitive to other absorbing gases in the sample cell.

Although these early gas-filter systems were utilized extensively in some commercial processes, their sensitivities and time responses are not sufficient for many modern applications.

Most modern nondispersive infrared absorption instruments are based on the negative-filtering method and use either the dual- or single-beam technique. A typical dual-beam design is illustrated in Fig. 3 and is seen to be similar to the early version of Fig. 2, with the exception that a single modern IR detector is now used, which requires chopping of the light beam. Sophisticated optics and an interference filter are also utilized to increase the instrument's sensitivity (Bartle et al., 1972). Furthermore, modern electronic processing must be used to convert the chopped detector ouput to a simple dc signal.

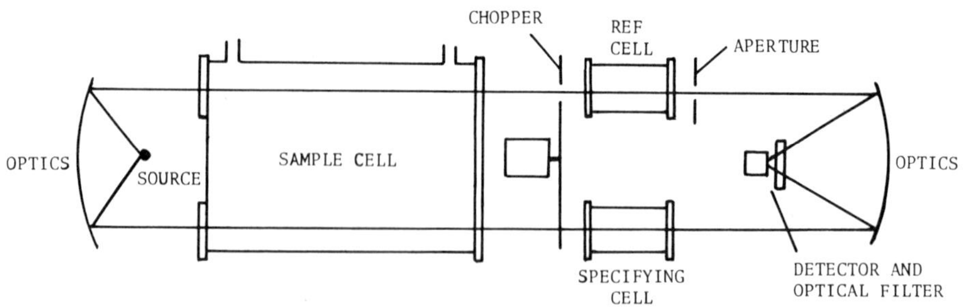

Fig. 3 Schematic of a modern dual-beam gas-filter analyzer.

The aperture shown in Fig. 3 is necessary to zero or balance the signal produced by the two light paths, since the beam through the specifying cell is significantly attenuated by the concentrated gas contained in that cell.

4 A MODERN SINGLE-BEAM GAS-FILTER INSTRUMENT

4.1 Optical Design

In order to eliminate dual-beam optical alignment problems and to simplify the zero balancing technique, the single-beam optical technique shown in Fig. 4 was developed (Sebacher, 1978).

The single-beam system includes (1) an IR light source; (2) a variable (1.5 to 50 m) multipass White cell (White, 1942), used as a sampling cell; (3) IR lens for focusing the beam; (4) a rotating cell partitioned so that half the cell acts as the specifying cell (filled with the gas to be measured) and the other half is filled with a gas that is optically transparent in the wavelength region of interest and acts as a reference cell; (5) a rotating chopper with nine evenly spaced slots on the reference side which are tapered to provide nulling of the electronic signal and ten evenly spaced untapered slots on the specifying-cell side (This chopper is permanently attached to the rotating cell so that the fast light chops will remain synchronized with the slower modulation chop from specifying cell to reference cell.); (6) a narrow IR bandpass filter; (7) a thermo-cooled Pb–Se IR detector; (8) reference-signal detectors; (9) a translational stage to move the rotating cell transverse to the IR beam; and (10) an electronic signal processor.

Infrared radiation from the source passes through the multipass sampling cell, where it is spectrally absorbed by the gas specified and other absorbing gases in the atmosphere. Upon leaving the sampling cell, the IR beam is focused by a lens onto a detector after being sent through a bandpass filter and alternately passed through the specifying cell and the reference cell and is chopped as the cell rotates. When the radiation passes through the reference cell, it is not attenuated further, but when it passes through the specifying cell, it is strongly attenuated in the spectral interval of the absorbing spectral lines. Therefore, a modulated chopped signal of the three frequencies is generated by the detector as shown in Fig. 4. As the concentration of the specified gas is increased in the sampling cell, the amplitude of the modulated signal (21 Hz) will change, and the

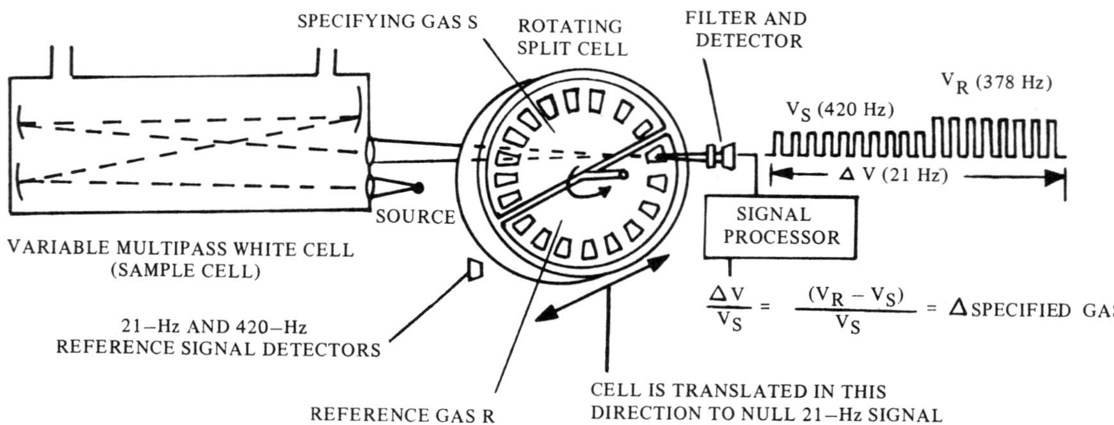

Fig. 4 Schematic of the single-beam gas-filter correlation instrument. V_S and V_R are the voltage outputs of the detector when the specifying cell or the reference cell is in the IR beam.

magnitude of this change in the 21-Hz signal is calibrated to the concentration of specified gas to be detected in the sampled atmosphere.

A narrow bandpass filter is selected which passes only a small portion of the absorption band of interest so that the detector views a very limited part of the wavelength interval being emitted by the source. In this manner the effects of interfering gases and source variations with wavelength are minimized.

To eliminate variations of the output signal due to source intensity fluctuations, the modulated signal change is electronically divided by the detector output when the specifying cell is in the light path. This signal, which does not vary with the specified gas concentration in the atmosphere, can be detected by a lock-in amplifier, since it has a chop frequency (420 Hz) different than the reference-cell chop (378 Hz).

4.2 Design Analysis

The operational requirements of the optical and electronic systems integrated into the gas-filter correlator can be further illustrated by the following simplified analysis. Using Beer's Law, the radiant power transmitted by a source is

$$P = \int P_0 \tau \, d\bar{\nu} \tag{3}$$

where $P_0(\bar{\nu})$ is the spectral radiant power from the source and $\tau(\bar{\nu})$ is the transmittance. When the cell is in the position in which the beam goes through the reference side, the radiant power reaching the detector is given by

$$P_R = \int \alpha_R P_0 \tau_G \tau_S' \, d\bar{\nu} \tag{4}$$

and when the cell turns so that the beam goes through the specifying side, the radiant power measured is

$$P_S = \int \alpha_S P_0 \tau_G \tau_S' \tau_S \, d\bar{\nu} \tag{5}$$

where α_R and α_S are the absorptances or attenuations of the beam from the optical parts of the reference cell and the specifying cell. τ_S is the transmittance due to the specifying gas in the cell, $\tau_{S'}$ is due to the specified gas in the atmosphere, and τ_G is the transmittance due to all other gases in the beam. Since the specifying-gas line frequency interval absorption in the specifying cell is much greater than in the atmosphere, the term $\tau_{S'}\tau_S$ in Eq. (5) can be replaced by τ_S, and τ_S is correlated with τ_S'. If a narrow bandpass filter is selected so that the detector only views a small portion of the wavelength interval being emitted, the effects of other absorbing gases are significantly reduced. Since only a very narrow wavelength interval is used, it will be assumed that Eqs. (4) and (5) can be replaced with average values of P_0, α, and τ so that

$$P_R = \bar{P}_0\bar{\alpha}_R\bar{\tau}_G\bar{\tau}_{S'} \tag{6}$$

$$P_S = \bar{P}_0\bar{\alpha}_S\bar{\tau}_G\bar{\tau}_S \tag{7}$$

If the beam, when passing through the reference side of the cell, can be attenuated, either optically or electronically, to balance the effects of absorption within the instrument so that $\bar{\alpha}_R = \bar{\alpha}_S\bar{\tau}_S$, then the differential voltage output is

$$\Delta V \propto \Delta P = P_S - P_R = \bar{P}_0\bar{\alpha}_S\bar{\tau}_G\bar{\tau}_S(1 - \bar{\tau}_{S'}) \tag{8}$$

To eliminate variations due to source changes, ΔV is divided by V_S, obtained from the detector output when the specifying cell is in the path. This signal, which is insensitive to variation of $\bar{\tau}_{S'}$, can be removed by a phase-sensitive detector (lock-in amplifier), since it has a chop frequency different than the reference-side chop. Then

$$\frac{\Delta V}{V_S} = \frac{P}{P_S} = \frac{\bar{P}_0\bar{\alpha}_S\bar{\tau}_G\bar{\tau}_S(1 - \bar{\tau}_{S'})}{\bar{P}_0\bar{\alpha}_S\bar{\tau}_G\bar{\tau}_S} \tag{9}$$

and

$$\frac{\Delta V}{V_S} = 1 - \tau_{S'} \tag{10}$$

The results of this analysis indicate that the instrument output is a function of the transmission of light through the specified gas.

4.3 Nulling the Output Signal

To obtain a zero output signal for zero specified gas concentration in the sampling cell, the 21-Hz signal must be nulled. Furthermore, the nulling out of the specifying-cell contribution to the 21-Hz signal allows one to increase the sensitivity of the electronic detection system for the measurement of very small 21-Hz signals resulting from very small gas concentrations. Two methods of nulling are possible with this system; one electronic and the other mechanical. When an electronic zeroing scheme is utilized, it is essential that the time constants for both the signal to be nulled and the nulling signal be as identical as possible. If the time constants are not matched, the 21-Hz electronic detection system will sense a phase shift when a fast fluctuation occurs that may be larger than the gas concentration signal being measured.

Since high-speed samplers usually require as fast a response as possible, a mechanical scheme for nulling the 21-Hz signal was devised which eliminates time constant matching in this part of the processing circuit. The mechanical nulling is incorporated into the chopping blade as illustrated in Fig. 5. Absorption of the IR beam by the gas in the specifying cell causes a lower voltage amplitude in that position of the chopped signal, as compared to the reference chop, and gives rise to the 21-Hz signal to be nulled. Since the electronic detection processing system must first demodulate the 21-Hz frequency from the three-frequency detector output, the rms values of the 420- and 378-Hz voltage signals will determine the amplitude of the 21-Hz signal. The rms value of the 378-Hz chop was decreased, as shown in Fig. 5, by tapering the reference-cell slots so that the pulse width could be varied by simply translating the chopping blade in a transverse direction to the IR beam. Translation of the chopper was accomplished by mounting the motor-driven rotating-cell assembly on a translational stage which could be controlled manually or electronically. In this manner the specifying-cell contribution to the 21-Hz signal could be removed without affecting the sampling-cell contribution.

4.4 Signal-Processing Circuits

A schematic of the electronic system designed for processing the IR detector output is shown in Fig. 6. Infrared radiation chopped at three frequencies [21 Hz (modulation chop), 420 Hz (specifying-cell chop), and 378 Hz (reference-cell chop)] falls on the Pb–Se detector. The output of the detector is amplified by a

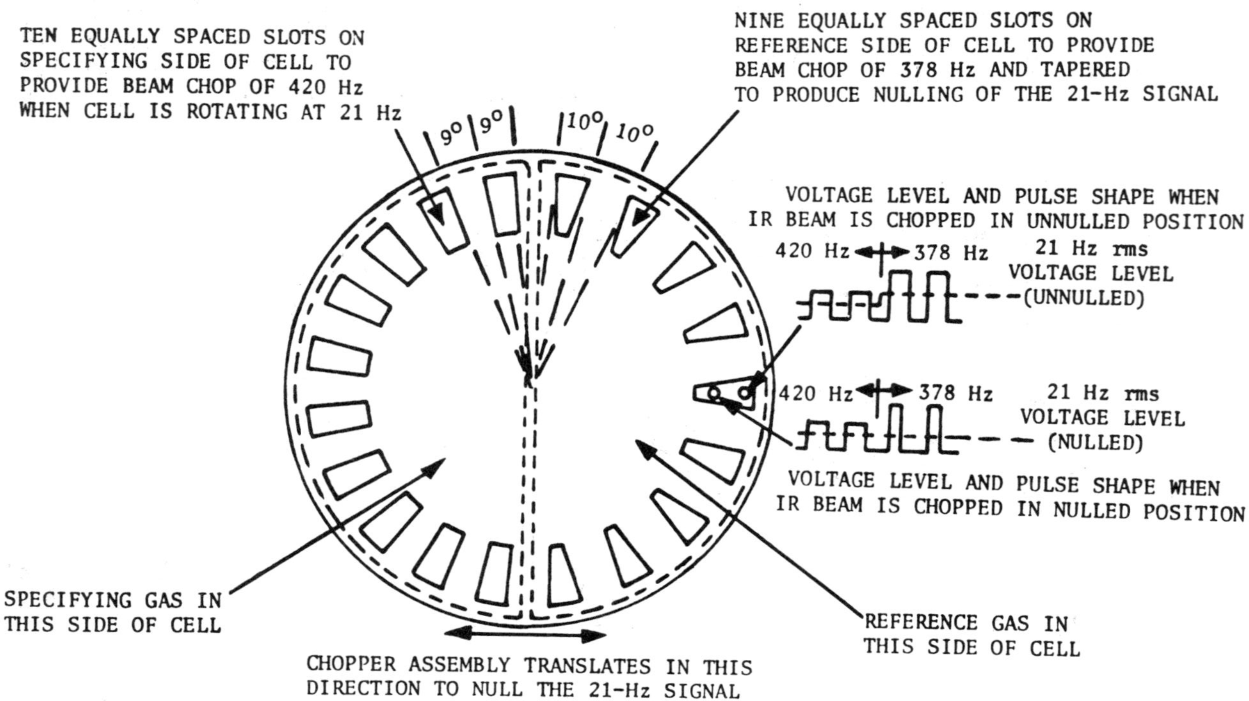

Fig. 5 Details of IR beam chopping blade which is permanently attached to the rotating split gas cell. Dashed line shows the outline of the split cell. IR windows seal the gas within the cell.

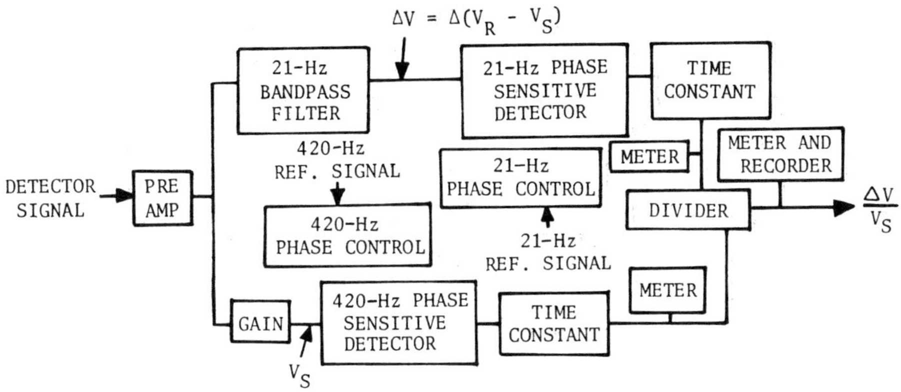

Fig. 6 Schematic of signal-processing circuit.

low-noise preamplifier located close to the detector to maintain the highest possible signal-to-noise ratio and then is split into inputs to 21- and 420-Hz circuit channels.

The 21-Hz input, which contains the information on variations in atmospheric specified gas concentration, is passed through a 21-Hz bandpass filter with variable gain to remove the 420- and 378-Hz components of the signal and to reduce the noise level. The output of the filter then goes to a 21-Hz phase-sensitive detector which is synchronized to the input signal by gate pulses from a 21-Hz phase control circuit which is derived from the 21-Hz reference-signal detector. The reference-signal detectors obtain their signals from separate choppers attached to the rotating cell. A phase-sensitive detector and phase-control circuit combination constitute a lock-in amplifier which produces a dc output that is proportional to the rms values of the signal input amplitude.

Demodulation of the 420-Hz signal is accomplished by sending the amplified IR detector pulse directly into a 420 Hz lock-in amplifier system which eliminates both the 21- and 378-Hz signals. The dc voltage output is therefore proportional to the input amplitude of the specifying-cell chopped signal. Range selection steps of input voltage to output voltage from 10^{-1} to 10^{-5} are provided to span a large range of gas concentrations. The phase-sensitive detectors are followed by low-pass filters with selectable time constants permitting longer integration times for low-level signals. The dual-channel dc signals are then fed into a divider circuit to produce the system output which is monitored by a strip recorder.

Panel meters are provided to read the output of each phase-sensitive detector and the divider output, and controls are available so that each component of the system is adjustable.

An alternative electronic processing scheme is shown in Fig. 7. In this circuit the 420-Hz circuit and divider have been replaced by an automatic gain control (AGC). Since any fluctuations of the light source will automatically be compensated for by the AGC, the fast chop peak input signal to the bandpass filter will stay constant and the 21-Hz signal will remain unaffected.

4.5 Spectral Characteristics Through Specifying Gas and Filter

Resolution of the gas filter is related to the linewidth of the molecular vibrational-rotational bands of the gas species to be measured and therefore depends on absorption line broadening. Because an optical bandpass filter is incorporated into the system, this technique uses a number of absorption lines which differ in linewidth. Since linewidth is a function of optical depth, the resolving power can be increased by decreasing either the gas concentration or optical length of the specifying cell.

Typical features of the characteristic transmission spectra are shown for three gases in Figs. 8, 9, and 10. These measurements were obtained by using an optical spectrograph which was scanned over the wavelength range of interest to measure the intensity of the IR beam after it passed through the interference filter and specifying gas. The CO absorption band is presented in Fig. 8 and includes part of the P and R branches of the CO band near 4.7 μm. The CH_4 absorption band near 3.3 μm is presented in Fig. 9 and includes P and R branch lines and the entire unresolved Q branch, which is not present in the CO and HCl bands. Characteristic features of the HCl band near 3.4 μm are seen in Fig. 10. This bandpass filter is slightly broader than the

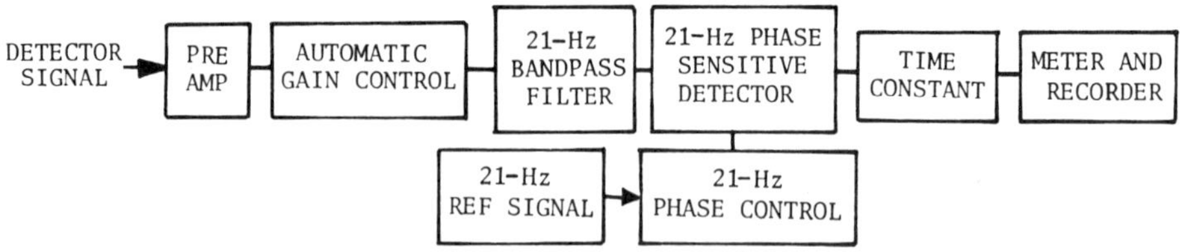

Fig. 7 Alternative signal-processing circuit.

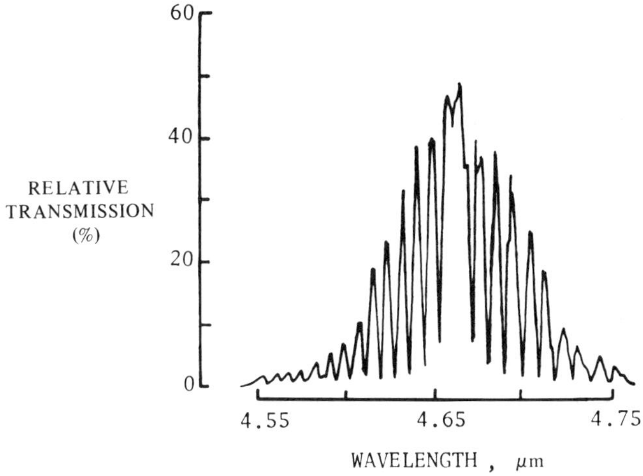

Fig. 8 Transmission of IR beam through 4.69-μm interference filter and CO specifying cell at band near 4.7 μm.

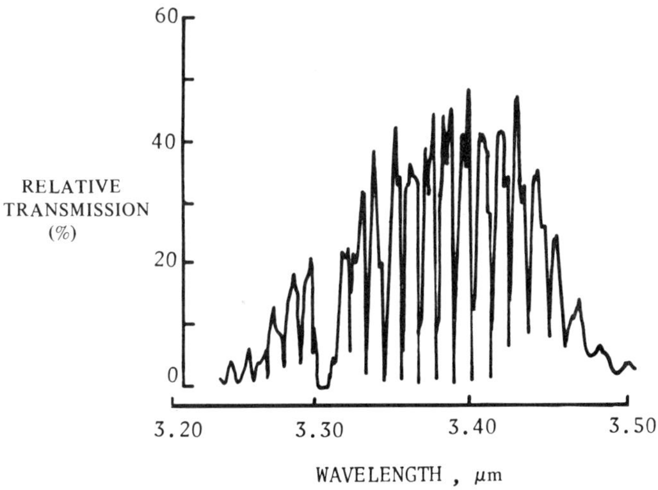

Fig. 9 Transmission of IR beam through 3.42-μm interference filter and CH_4 specifying cell at band near 3.3 μm.

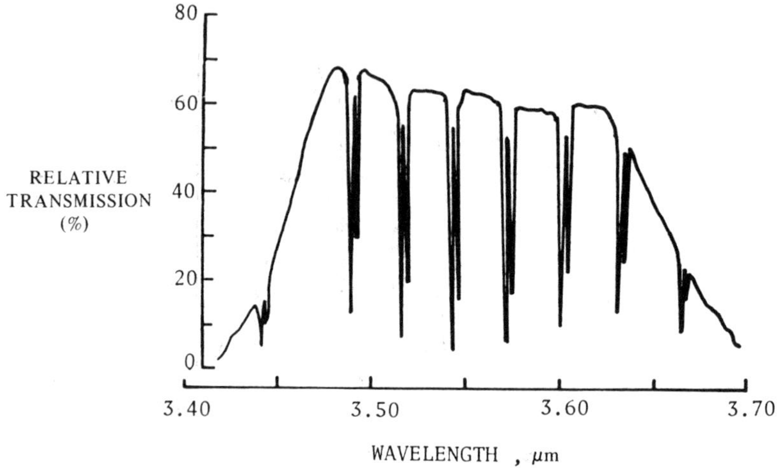

Fig. 10 Transmission of IR beam through 3.56-μm interference filter and HCl specifying cell at band near 3.4 μm.

others and includes only one R branch line and seven lines of the P branch. Doublets exist for each line of the band as a result of the isotopes $H^{35}Cl$ and $H^{37}Cl$. Undiluted instrument grade CO, CH_4, and HCl gases were used in the specifying cell; as a result, the optical thickness was 2.5 atm cm. Under these conditions, the radiation from the source is strongly absorbed at the characteristic wavelengths of each gas.

The line resolution shown in Figs. 8, 9, and 10 is a result of the spectrograph used and does not indicate the gas-filter correlator resolution, which is much greater. Resolution of the gas filter can be increased by decreasing the optical thickness of the specifying gas. Increased resolution becomes important if interfering gas absorption lines exist within the bandpass filter wavelength range and overlap the specifying-gas lines.

4.6 Calibration

To prepare the single-beam gas-filter correlation instrument shown in Fig. 4 for a particular gas, the specifying cell must first be charged with the gas to be measured and the proper interference filter installed in the detector assembly. The instrument is then electronically zeroed, and the gain controls are adjusted for a full-scale reading over the sensitivity range of interest. Gas mixtures containing known concentrations of the specified gas are then admitted to the sample cell for system calibration. Three typical calibration curves for CO, CH_4, and HCl are shown in Figs. 11, 12, and 13.

The lower detectable limit for the three gases shown is about 2 ppm-m and occurs when the signal-to-noise ratio reaches unity, a time constant of 1 sec being used. Actual gas concentrations as low as 0.02 ppm are measured by increasing the variable optical path length of the sampling chamber White cell to 50 m. Optical thickness values shown in the calibrations of Figs. 11, 12, and 13 must be divided by the preset optical path length to obtain the calibration for actual gas concentration. Lower detectable limits may be achieved by using longer time constants and by designing sampling chambers with longer optical path lengths. The upper detectable concentration is reached when the optical depth in the sampling path approaches the optical depth in the specifying cell. Measurements of trace gas concentrations above several hundred ppm are generally unnecessary in pollution measurements, but the calibrations indicate that the instrument is sensitive above 6000 ppm-m and is fairly linear up to 300 ppm-m.

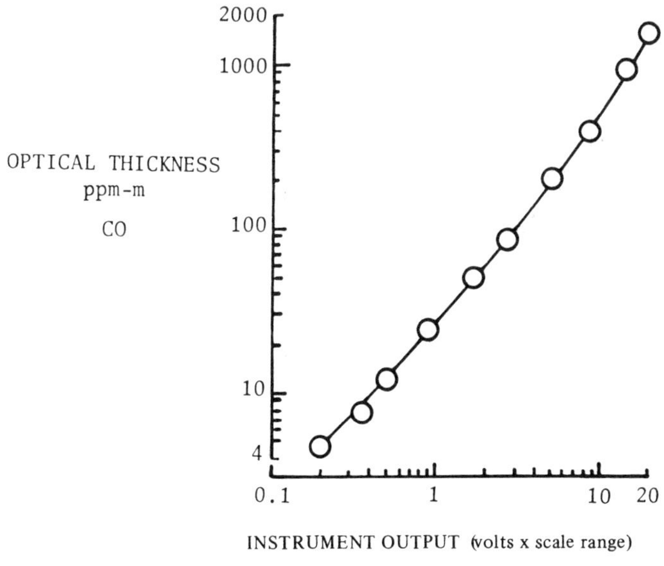

Fig. 11 Instrument calibration for CO at 1 atm.

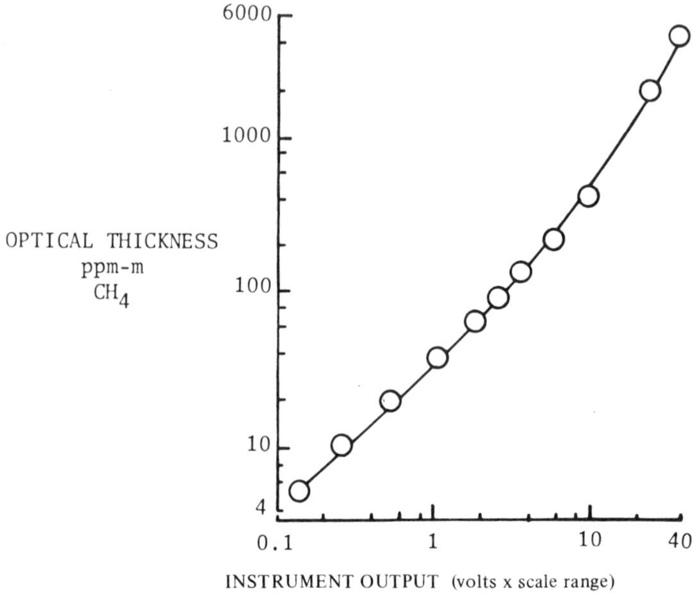

Fig. 12 Instrument calibration for CH_4 at 1 atm.

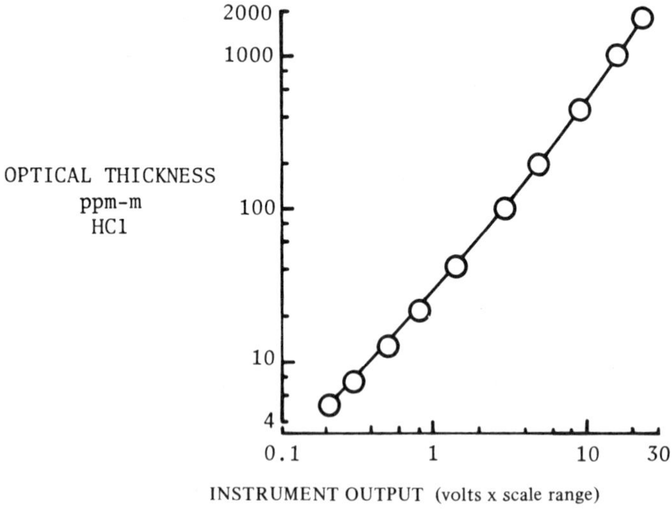

Fig. 13 Instrument calibration for HCl at 1 atm.

4.7 Interference Effects

Interfering effects on the CO, CH_4, and HCl systems were evaluated for other gases that may be present in the atmosphere, including CO_2, CO, H_2O, CH_4, C_2H_4, C_2H_6, C_3H_8, C_4H_{10}, and N_2O. These tests consisted of admitting known concentrations of the interfering gas into the calibration cell and observing instrument signals which would indicate equivalent concentrations of the specified gas. The results for CO are presented in Fig. 14, and only N_2O is found to cause interference. The CO calibration curve is also included in this plot for comparative purposes. Since a N_2O band is known to overlap the R branch of the CO band, this effect, although small, was expected. By using Figs. 11 and 14, it can be estimated that equivalent signals are obtained for about 1000 ppm-m of N_2O and 10 ppm-m of CO. If this ratio is extrapolated down to the anticipated midrange concentration for atmospheric CO, then 100 ppm of N_2O gives an equivalent signal as 1 ppm of CO. Since the measured atmospheric concentrations of N_2O are generally less than 500 ppb, the equivalent CO concentration would be less than 5 ppb. The resulting error would be insignificant for CO atmospheric measurements.

Results of the interfering-gas study for the CH_4 instrument version are shown in Fig. 15 along with the CH_4 calibration curve.

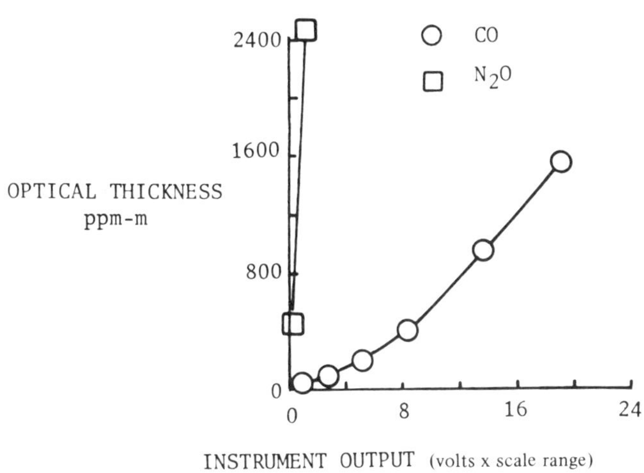

Fig. 14 A comparison of absorption interference effects of N_2O on CO gas-filter correlation analyzer with CO calibration.

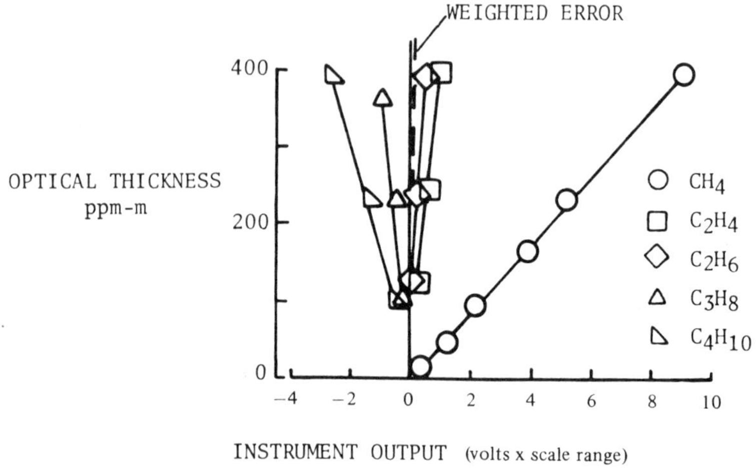

Fig. 15 A comparison of absorption interference effects of C_2H_4, C_2H_6, and C_4H_{10} on CH_4 gas-filter correlation analyzer with CH_4 calibration.

Positive interference effects were found for C_2H_4 and C_2H_6, but negative correlations were found for C_3H_8 and C_4H_{10}. Positive interference results from absorbing-gas lines that overlap the specified-gas lines in the filter wavelength interval and give rise to a decrease in the transmission through the reference cell but do not significantly affect the transmission through the specifying cell. Negative interference results from absorbing-gas lines that do not overlap the specifying-gas lines in the filter wavelength interval and are an effect of the nulling procedure. In this case, the transmission through both cells is decreased but by different amounts, since the zeroing technique has reduced the detector output when the reference cell is in the IR path in order to null the instrument output.

As in the CO instrument, it can be shown that the interfering effects of the hydrocarbons are small when the instrument is set up to measure atmospheric CH_4. Not only do the positive and negative interfering signals tend to cancel each other out, but the total hydrocarbon concentration of the atmosphere, excluding CH_4, is generally an order of magnitude less than the CH_4 concentration. Typical values measured (Warner, 1976) are as follows:

Gas	Value (ppm)
CH_4	2.355
C_2H_6	0.0636
C_2H_4	0.0636
C_4H_{10}	0.0276
C_3H_8	< 0.0001

If the interfering signal of each interfering hydrocarbon measured is weighted according to the ratio of that specific gas concentration to the CH_4 concentration listed above and then added, the total result will approximate the line of long dashes shown in Fig. 15. For a 2.4-ppm concentration of CH_4, the error would amount to about +1 percent or an equivalent CH_4 concentration of 24 ppb.

Although some overlap exists between the HCl and CH_4 bands, no CH_4 interference was found on the HCl version of the instrument. Of the gases tested, C_3H_8, C_2H_6, and C_4H_{10} did produce interference effects and all indicate a negative correlation as shown in Fig. 16. Again the results of these signals, when weighted according to their atmospheric concentration, would be insignificant for an HCl sensor used to measure this gas in the ppm range.

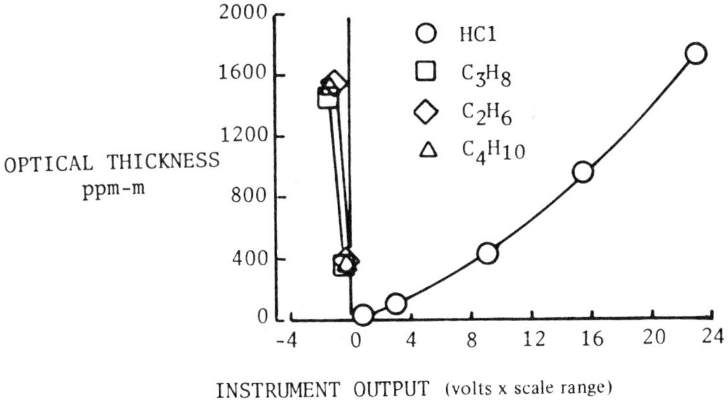

Fig. 16 A comparison of absorption interference effects of C_3H_8, C_2H_6, and C_4H_{10} on HCl gas-filter correlation analyzer with HCl calibration.

Interference effects may also occur from H_2O in the atmosphere. Since tropospherical moisture concentrations are highly variable, the most desirable requirement for any specifying gas is the ability to select an optical wavelength interval which contains no H_2O absorption lines. If this is not possible, a dessicant filter is usually placed on the intake line of the sampling chamber to remove moisture from the gas to be analyzed. When H_2O absorption lines interfere with a reactive gas, it may become necessary to monitor the calibrated H_2O response and subtract this output from the specified gas output.

5 APPLICATIONS

5.1 Industrial Gaseous Exhaust

Nondispersive infrared analyzers play a key role in monitoring modern industrial gaseous exhausts. These instruments are capable of sampling a stream continuously with excellent reliability, even under adverse conditions. Because of the adverse operating conditions which occur at plant sites and in industrial stack measurements, the sampling handling system becomes an important part of the analysis system. In these cases the sample temperature, pressure, solid content, and corrosiveness must be considered.

Most commercially available nondispersive plant stream analyzers make use of the negative filter or positive filter techniques. Since the total absorption technique does not discriminate between absorbing gases, it has not gained widespread acceptance. A number of companies manufacture a positive filter analyzer and usually use a capacitor microphone detector, although thermocouple detectors have been used. Negative filter plant stream analyzers are also commercially available, and many are based on the design of Fastie and Pfund (1947). Most commercial nondispersive instruments use relatively short path sample cells and have to be modified to allow for long-path or cross-stack measurements. In cross-stack measurements, an active IR source can still be used, while long-path measurements may be used against the sun or sky. Wiens and Zwick (1977) discuss the radiative characteristics of various natural backgrounds.

If cross-stack measurements in hot gases are desired, additional factors must be considered. First, hot gases are less dense than cold gases, so that for the same ppm-v in the hot gas, IR absorption and, therefore, detection is less sensitive than at ambient temperature. Second, the peak line absorption distribution within the band spectrum utilized by the instrument will change as a function of gas temperature. Third, as the temperature of the gas stream increases, spectral emissions from the hot gas will be sensed by the instrument detector at the same absorption wavelength used in the correlation technique. Since the detector cannot differentiate between stream emission and absorption of the source radiation, a room temperature calibration will be in error. Fourth, hot gases have different spectral linewidths than cold gases, so that the matching between the hot and calibration gaseous line shapes is not perfect. A detailed study on the temperature effects on line shape is to be found in Mallard and Gardiner (1978). Because of the difficulties involved in correcting the sensitivity changes resulting from the hot emitting gas environment, cross-stack measurements are usually avoided. Moderately high gas temperature conditions can be handled by calibrating the instrument at similar temperatures.

5.2 Gas-Flux Measurements

In addition to monitoring industrial gaseous exhausts and atmospheric trace gas concentrations, nondispersive infrared absorption techniques have been integrated into chamber systems for measuring gas fluxes from the air–soil and air–water interfaces. Modern gas-filter correlation instruments provide sufficient analytical sensitivity and rapid response time (less than 1 sec) to permit quantitative measurement of gas flux in measurement times of several minutes. In addition, modification of the standard flux chamber can provide the capability of quantitatively controlling air velocity over the water surface within the chamber sampling area, thereby providing a means of measuring gas flux as a function of air velocity (Sebacher and Harriss, 1982).

A schematic of a system used to measure CH_4 flux across the air–water interface as a function of air velocity, water temperature, and surface water-dissolved CH_4 is shown in Fig. 17. It consists of an open bottom chamber, partitioned into three sections. A lower section interfaces with the water surface and is constructed to provide a uniform airflow over the water surface by maintaining a constant cross-sectional area between airflow slots at either end of the lower section. A uniform flow is developed through the slots and across the water surface. The airflow is pumped by a blower into the larger sampling section and is then recirculated through the system. Air velocity is controlled by a valve incorporated into the blower. In addition, the sampling chamber air is circulated through the CH_4 analyzer and

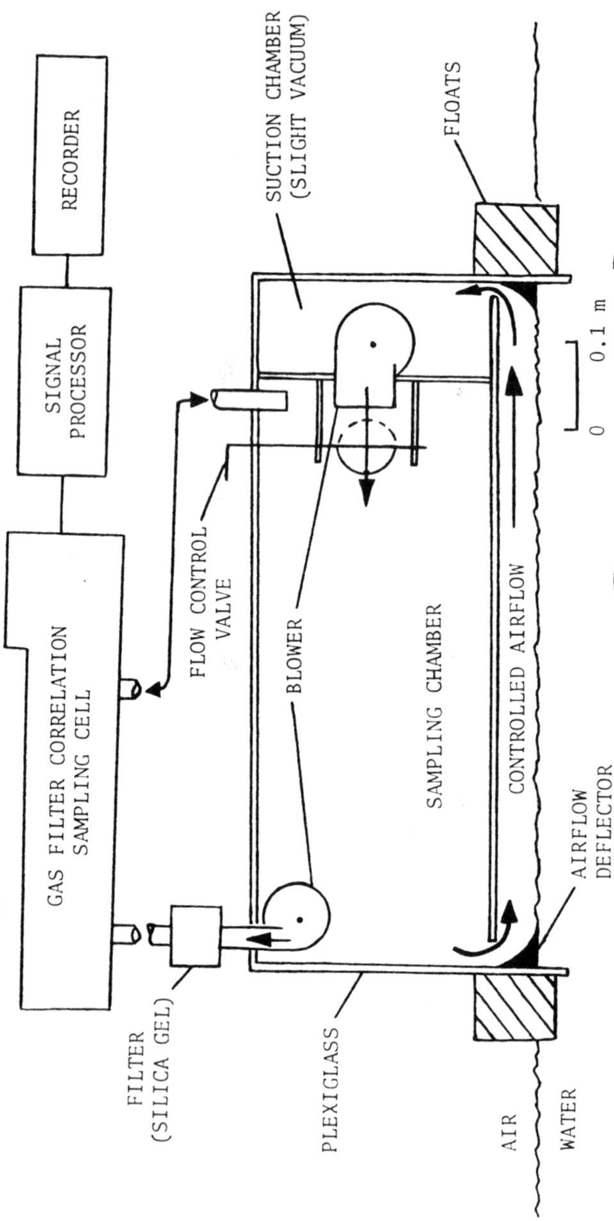

Fig. 17 System used to measure gas flux from a water surface as a function of air velocity.

back into the chamber so that any change in the CH_4 concentration in the chamber is continuously monitored. CH_4 flux is calculated directly from the recorded CH_4 concentration change with time, since the volume of the chamber and interface area are known. This system may be applied to any gas adaptable to nondispersive infrared absorption techniques and may be used to measure fluxes from industrial waste ponds as well as from natural sources.

Nondispersive absorption instruments have also been used to directly measure sea-air CO_2 flux by the eddy-correlation technique (Jones et al., 1978). When this method is applied to gas fluxes, it correlates fluctuations in the gas concentration at a fixed point with vertical wind fluctuations. Sensitivity and response time are the two main considerations for the instrument design. Response time of the sensor is determined by the turbulence of the air (between 0.02 and 10 Hz), and the sensitivity is determined by the magnitude of the concentration fluctuations resulting from the gas flux.

5.3 Satellite Observations

An important application of modern nondispersive infrared absorption techniques consists of large-scale regional mapping of atmospheric trace gases from space platforms. One instrument (MAPS) has recently been tested on an initial space shuttle flight and has been selected for other similar projects (Reichle et al., 1982; Hesketh et al., 1977). This passive remote sensor, which uses reflective natural sources of radiation from the Earth's surface, is illustrated in Fig. 18. The electro-optical head contains two gas cells, one at 266-Torr CO, the other at 76 Torr CO; a direct radiation detector; and external balance and gain check system; and an internal balance system. Thermal radiation passes up through the atmosphere into the viewport of the down-looking instrument. A beam of the incident radiation passes through the high pressure CO gas cell and onto a detector. This high pressure CO gas cell acts as a filter for the effects of CO present at low altitudes. A second beam falls directly onto a detector without passing through any gas filter to correlate the instrument for CO in the atmosphere at an altitude of 7–8 km. A third beam of the incident radiation passes through the low pressure CO gas cell and onto a detector. The low pressure CO gas cell filters out the effects of CO present at high altitudes. The difference in voltage from this and the direct detector provides a measure of CO concentration at an altitude of 10–12 km.

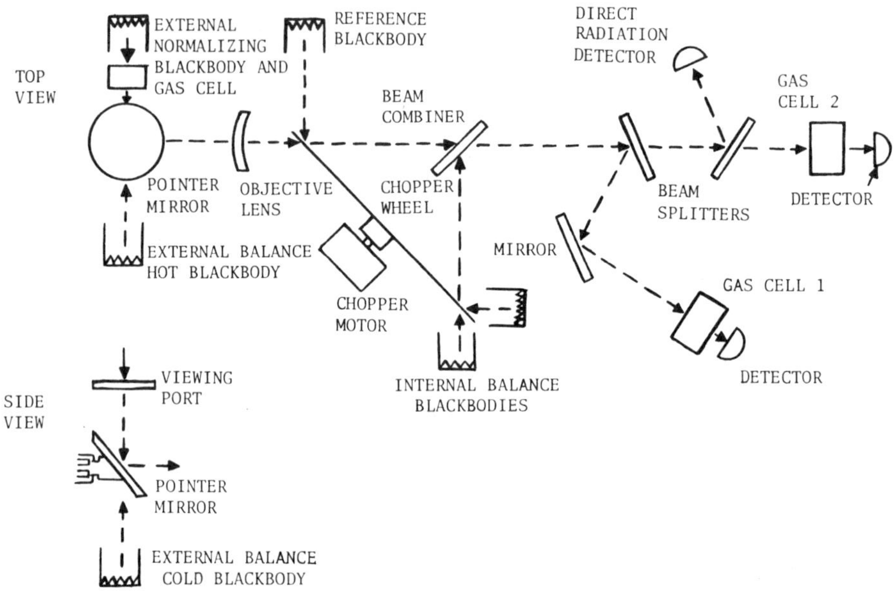

Fig. 18 Optical system layout for MAPS instrument.

The instrument is internally balanced so as to be relatively insensitive to changes in background temperature. Two blackbodies, one at the hot end of the expected temperature range, the other at the cold end, are introduced one at a time into the optical path with no gas between them and the instrument. The gain of each gas cell detector is adjusted so that each blackbody registers the same voltage, thus providing a baseline for signal measurement.

A similar system located externally serves to check both the balance set by the internal system and the instrument's gain stability. A pointer mirror rotates from a hot blackbody, to a cold blackbody, to a normalizing source, and finally to the viewport. Data-taking continues throughout the Earth-observing period, with balance and gain check recurring at 12-h intervals or upon command.

REFERENCES

Bartle, E. R., Kaye, S., and Meckstroth, E. A. (1972). An in-situ monitor for HCl and HF, *J. Spacecraft and Rockets*, 9: 836.

Davies, J. H., Barringer, A. R., and Dick, R. (1982). Gaseous correlation spectrometric measurements, in *Digest of Technical Papers Presented at the Workshop on Optical and Laser Remote Sensing*, 9–11 February 1982, Monterey, CA, sponsored by U.S. Army Research Office.

Fastie, W. G., and Pfund, A. H. (1947). Selective infra-red gas analyzers, *J. Opt. Soc. Am.*, *37*: 762.

Fowler, R. C. (1949). A rapid infra-red analyzer, *Rev. Sci. Instrum.*, *20*: 175.

Hanst, P. L., Lefohn, A. S., and Gay, B. W. (1973). Detection of atmospheric pollutants at parts-per-billion levels by infrared spectroscopy, *Appl. Spectrosc.*, *27*: 188.

Herzberg, G. (1950). *Molecular Spectra and Molecular Structure, I. Spectra Diatomic Molecules*, Second ed., D. Van Nostrand Co., Inc, New York.

Hesketh, W. D., Reichle, H. G., Massey, W. A., Ward, T. V., and Zwick, H. H. (1977). In *Remote Sensing of Earth's Resources* (V. F. Shahrakhi, ed.), Univ. of Tenn. Space Institute, Tullahoma, 527.

Hodgeson, J. A., McClenny, W. A., and Hanst, P. L. (1973). Air pollution monitoring by advanced spectroscopic techniques, *Science, 182*: 248.

Jones, E. P., Ward, T. V., and Zwick, H. H. (1978). A fast response atmospheric carbon dioxide sensor for eddy correlation flux measurements, *Atm. Environ.*, *12*: 845.

Mallard, W. G., and Gardiner, W. C. (1978). Absorption of the 3.39 μm He-Ne laser line by methane from 300 to 2400 K, *J. Quant. Spectrosc. Radiat. Transfer, 20*: 135.

Reichle, H. G., Beck, S. M., Haynes, R. E., Hesketh, W. D., Holland, J. A., Hypes, W. D., Orr, H. D., Sherrill, R. T., Wallio, H. A., Casas, J. C., Saylor, M. S., and Gormsen, B. B. (1982). Carbon monoxide measurements in the troposphere, *Science, 218*: 1024.

Sebacher, D. I. (1978). Airborne nondispersive infrared monitor for atmospheric trace gases, *Rev. Sci. Instrum.*, *49*: 1520.

Sebacher, D. I., and Harriss, R. C. (1982). A system for measuring methane fluxes from inland and coastal wetland environments, *J. Environ. Qual.*, *11*: 34.

Ward, T. V., and Zwick, H. H. (1975). Gas cell correlation spectrometer: GASPEC, *Appl. Opt.*, *14*: 2896.

Warner, P. O. (1976). *Analysis of Air Pollutants*, John Wiley and Sons Inc., New York, p.

White, J. U. (1942). Long optical paths of large aperture, *J. Opt. Soc. Am.*, *32*: 285.

Wiens, R. H., and Zwick, H. H. (1977). Trace gas detection by correlation spectroscopy, in *Infrared, Correlation, and Fourier Transform Spectrometry* (J. S. Mattson, 11. B. Mark, Jr., and H. C. MacDonald, Jr., eds.), Marcel Dekker, Inc., New York, New York.

Wright, N., and Herscher, L. W. (1946). Recording infrared analyzers for butadiene and styrene plant streams, *J. Opt. Soc. Am.*, *36*: 195.

Appendix

AVAILABILITY OF COMPUTER CODES

This appendix provides information on how to obtain computer codes mentioned in the text from appropriate government agencies. Privately written codes can often be obtained by contacting the authors of the referenced publication describing the codes. For codes like the Air Force Geophysics Laboratory atmospheric transmission programs FASCOD (used in Chap. 4) and LOWTRAN as well as the AFGL absorption line listings for major and trace atmospheric species (see Chap. 1), which have been through multiple revisions and are included in an ongoing program of development and support, distribution has been institutionalized. Magnetic tapes are available, upon payment and specification of computer format, from

National Climatic Center, NOAA
Environmental Data Services
Federal Building
Asheville, NC 28801

The present technical point of contact, responsible for example for notifying users of changes in the codes, is

F. X. Kneizys
AFGL/OPI
Hanscom AFB, MA 01731

The industry standard plume flow-field and infrared radiation codes (SPF and SIRRM) described in Chap. 1 were developed by the Exhaust Plume Technology Subcommittee of the Joint Army-Navy-NASA-Air Force Interagency Propulsion Committee, known as JANNAF. Information exchange for JANNAF is handled by the Chemical Propulsion Information Agency (CPIA), whose address, including the subcommittee CPIA representative, is

The Johns Hopkins University
Applied Physics Laboratory
Attn: CPIA, T. M. Gilliland
Johns Hopkins Road
Laurel, MD 20707

Codes are not available through the CPIA, only information on where to obtain them. At present, the codes have only been released for use on specific government contracts, and code distribution is still handled by technical personnel in the lead developing organizations. For SPF, the appropriate point of contact is

Commander, U.S. Army Missile Command
Attn: DRSMI-RDK, E. Vaughn (or Dr. B. Walker)
Redstone Arsenal, AL 35809

The developing organization and source of manuals and tapes for SIRRM is

Air Force Rocket Propulsion Laboratory
Attn: AFRPL/DY, Dr. P. Kessel
Edwards AFB, CA 93523

although to date most tape distribution has also been done by the Army.

The above information is given with the knowledge that the continually changing government organization and the often short tenure of government points of contact may quickly render it obsolete. Giving both an institutional and a technical address is only intended to slow down that process—eventually, prospective users may begin with the above addresses and still invest considerable detective work.

Index

I

J

L

M